The Politics of Climate Change Negotiations

NEW HORIZONS IN ENVIRONMENTAL POLITICS

Series Editor: Arthur Mol, *Chair and Professor in Environmental Policy, Director, Wageningen School of Social Sciences, Wageningen University, The Netherlands, Professor in Environmental Policy, Renmin University, Beijing*

The New Horizons in Environmental Politics series provides a platform for in-depth critical assessments of how we understand the many changes in the politics of nature, the environment and natural resources that have occurred over the last fifty years. Books in the series question how the environment is (re)defined, debated and protected; explore differences between countries and regions in environmental politics; analyse how actors do and do not collaborate around environment and natural resource conflicts; describe who wins and who loses and in what ways; and detail how to better study, analyze and theorize such developments and outcomes.

The series is designed to promote innovative cross-disciplinary analysis of the contemporary issues and debates influencing the various dimensions of environmental politics. Covering a diverse range of topics, the series will examine the political, economic and ethical aspects of environmental policy, governance and regulation. It brings together cutting edge research on environmental politics worldwide in order to shed light on, and explain current trends and developments.

With oversight from the Series Editor, Professor Arthur Mol – a noted specialist in the field of environmental politics at Wageningen University, The Netherlands – the New Horizons in Environmental Politics series comprises carefully commissioned projects from experts in the field including both academics and professionals. The audience for the series is global, and books in the series are essential reading for students, academics and professionals, in short, anyone with an interest in understanding the vital issues affecting environmental politics in the 21st Century.

Recent titles in the series include:

Subnational Partnerships for Sustainable Development
Transatlantic Cooperation between the United States and Germany
Holley Andrea Ralston

The Politics of Climate Change Negotiations
Strategies and Variables in Prolonged International Negotiations
Christian Downie

The Politics of Climate Change Negotiations

Strategies and Variables in Prolonged International Negotiations

Christian Downie

Regulatory Institutions Network, The Australian National University

NEW HORIZONS IN ENVIRONMENTAL POLITICS

Edward Elgar

Cheltenham, UK • Northampton, MA, USA

Published by
Edward Elgar Publishing Limited
The Lypiatts
15 Lansdown Road
Cheltenham
Glos GL50 2JA
UK

Edward Elgar Publishing, Inc.
William Pratt House
9 Dewey Court
Northampton
Massachusetts 01060
USA

A catalogue record for this book
is available from the British Library

Library of Congress Control Number: 2013949884

This book is available electronically in the ElgarOnline.com Economics Subject Collection, E-ISBN 978 1 78347 211 6

MIX
Paper from responsible sources
FSC® C013056
FSC
www.fsc.org

ISBN 978 1 78347 210 9

Typeset by Columns Design XML Ltd, Reading
Printed and bound in Great Britain by T.J. International Ltd, Padstow

Contents

Acknowledgements

The process of writing this book has not been a linear one, and like negotiations, many individuals have been crucial to ensuring that this study reached a successful conclusion. Two people have done more than anyone to make this happen; Professor Neil Gunningham and Professor Peter Drahos. I could not have asked, or even hoped, for more from my thesis supervisors, from which this book originated. As I sought out supervisors I was told that it is the most important academic decision one makes but: 'you'll need some luck'. I have been very lucky. I am especially grateful for their academic guidance, confident encouragement and sense of humour. In truth, I could not have written this book without them, it has been a pleasure taking this academic journey in their company. Their supervision has been complemented by two other academics at the Australian National University (ANU), Dr Kyla Tienhaara and Professor Stephen Howes. Kyla began as a friend and later became a supervisor. She has been terrific in both roles. I am especially thankful for her detailed reading of each and every page of my final draft. Stephen kindly shared his economic insights with me as I pondered 'why states make the decisions they do'. Economics is more often than not a factor, and he helped me to see how.

The efforts of Americans and Europeans have been crucial to determining the outcome of many rounds of the climate negotiations, and they have also been crucial to shaping my research. One person stands out; Professor Lawrence Susskind. In 2010, I had the opportunity of working with him at the Massachusetts Institute of Technology. I am indebted to him not only for inviting me to stay at MIT, but for the countless hours he spent challenging my empirical findings and theoretical conclusions. These conversations provided an invaluable intellectual stimulus as I finalised the structure and argument of the thesis. I also had the pleasure of discussing my research with many other academics in the United States who I had long admired from afar. In particular, I am grateful to Professors Daniel Druckman, Adil Najam, Robert Putnam and William Zartman.

As I set out on this study, one factor that was completely outside my control was informants. Elite interviewing is not an easy task and its success depends on the generosity of others to give of their time, their knowledge and often their patience. I cannot thank each and every individual I interviewed on both sides of the Atlantic, and many would prefer to remain anonymous. However, several people gave more than others. First and foremost, Nigel Purvis, who I first interviewed in 2010 and later worked alongside at Resources For the Future in Washington DC in 2011. Nigel made this and much more happen. Many others not only agreed to be interviewed but kindly read sections of the manuscript providing invaluable feedback and often correcting my historical inaccuracies, including Marius Enthoven, Professor Jeffrey Frankel, Jørgen Henningsen and Dr Bert Metz.

In 2011 I returned to ANU to begin furiously drafting and re-drafting each chapter. Several people aided this process considerably. Adjunct Professor Howard Bamsey generously edited parts of the thesis and as a former negotiator he has been a wonderful source of knowledge on the history of the climate negotiations. In addition, Professors Carsten Daugbjerg and John Ravenhill both read and commented on draft chapters. Similarly, I should also acknowledge Professor Clive Hamilton who I first worked with in 2006, and in many ways his passion for academic research and public policy, especially in the area of climate change, ultimately informed my own decision to write a book in this field.

When I say I returned to ANU, I am referring to a particularly special environment; the Regulatory Institutions Network (RegNet) is a wonderful place to work. As an academic community it is unique not only for its prodigious output of world class research, but for the warmth and generosity of many people in the hallways who are always willing to assist with the personal and academic trials of research. In particular, I would like to thank Professors John Braithwaite and Valerie Braithwaite for their ideas and encouragement, especially in the early stages of my research. Also, to the many past and present PhD students who have been a source of strength and support along the way, including Dr Russell Brewer. This process has been a lot easier in your company. I would also like to thank the fantastic team at Edward Elgar who have been a pleasure to work with.

Finally, a thesis after all, is only that. While I may have lived and breathed this research over the last few years, I have had the good fortune of spending my time away from it in the company of many loving people. Most of all to Yolanda, Andrew, John, Brett and Amy for making this and everything else I do possible. And, to a brilliant and diverse circle of friends in Australia and overseas, many who supported me on my travels, it was, and continues to be, a pleasure. I am indebted to each of you.

Abbreviations

AGBM	Ad Hoc Group to the Berlin Mandate
AOSIS	Association of Small Island States
API	American Petroleum Institute
BMU	Federal Ministry for the Environment, Nature Conservation and Nuclear Safety
BMWi	Federal Ministry of Economics
BP	British Petroleum
BTU	British Thermal Units
CAN	Climate Action Network
CBI	Confederation of British Industry
CCAP	Center for Clean Air Policy
CDM	Clean Development Mechanism
CEA	Council of Economic Advisors
CEFIC	European Chemical Industry Council
CEQ	Council of Environmental Quality
COG	Chief of Government
COP	Conference of the Parties
DETR	Department of the Environment, Transport and the Regions
DG	Director-General
DoE	Department of Energy
EC	European Commission
EDF	Environmental Defense Fund
EIA	Energy Information Administration
EPA	Environmental Protection Agency
EU	European Union
EURELECTRIC	Union of the Electricity Industry
EUROFER	European Confederation of Iron and Steel Industries
EUROPIA	European Petroleum Industry Association
FIELD	Foundation for International Environmental Law and Development
G77	Group of 77
GCC	Global Climate Coalition
IAT	Interagency Analytical Team

ICC	International Chamber of Commerce
INC	Intergovernmental Negotiating Committee
IPCC	Intergovernmental Panel on Climate Change
IUCN	International Union for Conservation of Nature
MOP	Meeting of the Parties
NEC	National Economic Council
NRDC	Natural Resources Defense Council
NSC	National Security Council
POMs	Policies and Measures
RSPB	Royal Society for the Protection of Birds
TRIPS	Trade-related Aspects of Intellectual Property Rights
UN	United Nations
UNFCCC	United Nations Framework Convention on Climate Change
UNICE	Union of Industrial and Employers' Confederations of Europe
US	United States
VROM	Ministry of Housing, Spatial Planning and the Environment
WCI	World Coal Institute
WWF	World Wide Fund for Nature

PART I

International negotiations and theoretical background

1. Introduction

THE PUZZLE OF PROLONGED INTERNATIONAL NEGOTIATIONS

International negotiations matter in world politics. Questions of international security, trade and the environment cannot be addressed if states do not engage in international negotiations. From the Paris and Potsdam peace treaty negotiations in the first half of the 20th Century to the ozone and climate change negotiations in the second, the last 100 years has been marked by the success or failure of international negotiations and the 21st Century looks set to be the same. Many of these negotiations, indeed many of the most significant in the post-war era, have been prolonged, stretching for years and sometimes decades. This has certainly been true for environmental and trade issues. The international negotiations on the Law of the Sea and the Uruguay Round of trade negotiations both lasted almost a decade and the most recent Doha Round of trade negotiations will surpass them both.

The international climate negotiations represent one of the best examples of the phenomenon. Since the Rio Earth Summit in 1992 and the entry into force of the United Nations Framework Convention on Climate Change (UNFCCC) in March 1994, states have been engaged in a seemingly endless negotiation process. The only constant has been the continuation of the process, punctuated by agreement and acrimony. The so-called 'Kyoto phase' of negotiations, which commenced with the first Conference of the Parties to the UNFCCC in 1995, and took a decade to conclude before the Kyoto Protocol ultimately entered into force at the eleventh Conference of the Parties in 2005. Since then, negotiations have commenced on a 'post-Kyoto phase', which threatens to be as protracted as the last round.

Although prolonged international negotiations, such as the climate negotiations, seek to address some of the most critical problems facing the globe, the phenomenon is not well understood. Why does this matter? It matters because the less we know about prolonged international negotiations the less likely states are to reach agreement, and without agreement in such negotiations the world will get warmer, trade talks will stall and future negotiations around food, water or security are more likely to end in impasse. In short, the future of humanity will be at risk should some of the most significant current and future prolonged international negotiations fail.

That is not to say that international negotiations have been ignored, they have not. In the post-war era, international negotiations have been an important focus of study in the social sciences. Most of the literature has been concerned with explaining why states cooperate. In other words, how and why do some international negotiations end in agreement and others impasse? Yet remarkably almost none of this work has considered these questions for prolonged international negotiations. While great strides have been made in improving our understanding of international negotiation outcomes by examining the negotiation process, and/or the larger context surrounding the negotiation (see for example, Susskind and Crump, 2008: Vol. 4), theorizing has not considered these factors when negotiations become protracted. For example, extensive work has been done on the role of state and non-state actors in international negotiations, on the influence of domestic pressures and domestic political institutions (Putnam, 1988; Moravcsik, 1993), on the role of transnational activities of state and non-state actors (Keohane and Nye, 1972; Keohane and Nye, 1974; Risse-Kappen, 1995a; Milner and Moravcsik, 2009; Downie 2013) and on the impact of international regimes (Krasner, 1983; Levy et al., 1995; Young, 1999). Yet very little work has been undertaken on how these factors vary over time in negotiations. Most of this work emphasizes the negotiation outcome, or end game. The focus is on the actors in the end game. Who they were? What they did? For instance, if one is to inquire into how a specific state actor, such as the chief of government, or a non-state actor, such as an environmental NGO, influenced the likelihood of states reaching an agreement there is an abundance of studies to draw on. However, if one is curious about how the actions of either of these actors varied over time and the impact this had on a prolonged international negotiation there is a notable absence of empirics or theory. Consequently, the focus here is on the *temporal dimension*. It is not simply the end game, or the actors, or a class of actors, the focus is on actors over time. Instead of looking at each individual negotiation as one would a still photo, the aim here is to put them together to create a moving picture so we can analyse variations in the behaviour of actors over the course of an extended negotiation.

The central research questions then, are:

In prolonged international negotiations, what factors lead a state to change its negotiating position and the type of agreement it is willing to sign? And, how and why are these decisions made?

What is theoretically distinct about these questions is that they are asked in the context of *prolonged international negotiations*. Thus far, I have simply referred to prolonged international negotiations as international negotiations that continue for years and sometimes decades. More specifically, prolonged international negotiations can be defined as substantive international negotiations over a legally binding instrument that continue for five or more years, which begin with bargaining over a tentative agreement and conclude with bargaining over the ratification of that instrument. While the rationale for this definition will be elaborated in the next chapter, suffice to say that this definition can encompass the negotiation of consecutive non-binding agreements on the path towards a legal instrument. However, it does not include the ongoing monitoring, reporting, verification and enforcement processes that follow ratification, which proceed for years, and often decades, as parties meet to review the implementation of a legally binding instrument.

In order to address the central research questions, it is imperative that we better understand prolonged international negotiations. In particular, that we understand how and why states behave in international negotiations that continue for years and sometimes decades and the impact this has on the likelihood of an agreement being reached. This book takes up this task by analysing the behaviour of the United States (US) and the European Union (EU) in the international climate negotiations. The climate negotiations may not necessarily represent a trend toward more protracted international negotiations, but they do provide a good example of this phenomenon. And, because they seek to address one of the most pressing problems facing the globe, the climate negotiations on their own are especially significant.

CLIMATE CHANGE AND THE INTERNATIONAL NEGOTIATIONS

The Climate Change Problem

There is now scientific consensus that human-induced climate change is happening and that it threatens our planet. If the world does not take action to reduce global greenhouse gas emissions it is projected that by the end of the century average global temperatures will rise by 5°C above pre-industrial levels. The results would be horrific. Mass extinctions, large scale melting of the Greenland ice-sheet, accelerated disintegration of the West Antarctic ice sheet and increased frequency and severity of storms and floods on the one hand, and droughts and fires on the other. It is generally accepted that for some of the worst risks of climate change to be avoided, the average rise in global temperatures must be kept to no more than 2°C above pre-industrial levels (IPCC, 2013; Allison et al., 2011). For a 2°C target to be met, or any target for that matter, the world must agree to reduce emissions and that requires international negotiations.

However, the problem of climate change presents international negotiators with a unique set of challenges, which is why it is often described as a 'malign' or 'wicked' problem (Depledge, 2005: 18; Garnaut, 2008). First, climate change is truly a global problem. Greenhouse gases are well mixed in the atmosphere, which means there is no relation between where the gases are emitted and where the effects are felt. It also means that every country in the world has a stake in the outcome, even those that contribute very little to the problem. Crucially, every country will also benefit from the contribution of others in reducing greenhouse gas emissions even if they do not contribute themselves. In other words, climate change is a problem of global collective action.

Second, the causes of climate change – fuel combustion and land-use change – are associated with almost every human activity. As a result, it is not only states who are concerned about climate change but industry, environmentalists, farmers and indigenous groups, among others. Almost every group in society can be said to have a stake in the problem of climate change. This implies that negotiations have to address the concerns of a multitude of actors (Porter et al., 2000; Depledge, 2005). Third, given the causes of climate change, for many countries it is not an environmental issue, but an issue of development. For most industrialized countries the path to development has been carbon intensive, that is, they have become rich by burning fossil fuels. Developing countries therefore have long viewed any restrictions on the burning of fossil fuels as a restriction on their development. This has created great friction between developed and developing countries, one that has not been limited to the climate change negotiations (Susskind, 1994: Ch 1).

Fourth, climate change is also an especially malign problem because it is characterized by uncertainty. Scientifically, the climate system is immensely complex and the concentration of greenhouse gas emissions is one of many factors (others include solar radiation and ocean currents) determining global temperatures. As a result, it is often difficult to isolate human-induced climate changes from natural variability and to predict the precise long-term impacts of changes in the climate (IPCC, 2013). This in turn creates economic uncertainty in responding to the problem. Decision makers informed by economists have found it difficult to account for the long time horizons of climate change, which requires current generations to shoulder short-term costs, whose benefits will only be experienced by future generations (Stern, 2007).

Finally, failure to address the problem of climate change risks 'abrupt or irreversible changes' in the earth's climate (IPCC, 2013: 53). For example, it is predicted that the loss of ice sheets could lead to multi-metre sea level rise, which would permanently shift coastlines. Mass extinctions are also likely as global temperatures rise. The irreversible and abrupt nature of these changes means that increasing greenhouse gas emissions could shift the climate into a different state in a short period of time. To avoid these risks climate change must be addressed urgently. Scientists have concluded that the rise in global greenhouse gas emissions needs to be stabilized by 2015 if a 2°C target is to be met (IPCC, 2013: 67). The time imperative is one more reason why climate change poses a unique set of challenges for international negotiations.

An Overview of the International Climate Change Negotiations

Despite the wicked nature of the problem, states have attempted to address it and more than two decades of international climate negotiations have taken place – see Table 1.1. While many lament their failure to reduce global greenhouse gas emissions, the negotiations have produced some remarkable successes, which seemed unlikely at the time, including the UNFCCC in 1992 and the Kyoto Protocol in 1997 (Grubb et al., 1993; Oberthur and Ott, 1999).

Climate change first appeared on the international stage in September 1988, when Malta requested the UN General Assembly to include an agenda item on the topic. The debate in the General Assembly followed the discovery of the hole in the ozone layer and the successful adoption of the Montreal Protocol on Substances that Deplete the Ozone Layer in 1987. Yet 1988 was the critical year for the history of climate change. The Intergovernmental Panel on Climate Change (IPCC) was established, the Toronto Conference on the Changing Atmosphere took place and an extraordinary heat wave in the US occurred. NASA scientist James Hansen famously stated at a congressional inquiry in that year, that 'the earth is warmer in 1988 than at any time in the history of instrumental measurements' (Bodansky, 1993). As the *New York Times* declared the next day, 'Global warming has begun' (Shabecoff, 1988).

While climate change was by no means a mainstream issue, these events helped to set the stage for the formal international negotiations. Using the above definition of a prolonged international negotiation, which refers to substantive negotiations over a legal instrument, the international climate negotiations can be divided into three phases: the UNFCCC phase from 1991 to 1994, the Kyoto phase from 1995 to 2005, and the post-Kyoto phase from 2006 to present. The UNFCCC phase of negotiations formally commenced in February 1991 with the first meeting of the Intergovernmental Negotiating Committee (INC) for a Framework Convention on Climate Change. In the lead up to the Rio Earth Summit, the INC met five times as it worked to finalize text for the UNFCCC. After much contention between the US and the EU over binding commitments to limit greenhouse gas emissions, which were ultimately left out after US President George Bush opposed any reference to such commitments, the UNFCCC was signed by 154 countries including the US in Rio in June 1992 (Bodansky, 1993; Grubb et al., 1993; Porter et al., 2000: 117). In the words of the UNFCCC, the 'ultimate objective' is 'the stabilization of greenhouse gas concentrations in the atmosphere at a level that would prevent dangerous anthropogenic interference in the climate system' (UNFCCC, 1992). The signing of the Convention and its rapid entry into force less than two years later in March 1994, after ratification by the minimum necessary fifty states, marked the end of the first phase of the international climate negotiations.

After the triumph of the UNFCCC, the first Conference of the Parties (COP) to the UNFCCC held in Berlin in March 1995 signalled the beginning of the Kyoto phase of negotiations. This second phase was to prove one of the most intensive periods of negotiations, especially as the parties negotiated the Kyoto Protocol. At COP 1 the challenge was to decide whether the commitments of developed countries were 'adequate' to meet the ultimate objective of the UNFCCC to prevent dangerous climate change (UNFCCC, 1992). The end result was the Berlin Mandate, which explicitly recognized that they were not, and that work should begin toward a protocol or other legal instrument to be completed in 1997. The Kyoto Protocol, which was finalized at COP 3 in December 1997, was a landmark agreement. It introduced binding emission targets for all developed countries, something that the US had been adamantly opposed to, and a series of flexibility mechanisms to assist countries to meet their targets, something that the EU had viewed as a loophole for the US and others (Oberthur and Ott, 1999: Ch 7).

However, COP 3 in Kyoto did not finalize the rules of the Protocol and negotiations floundered until 2001 when the Marrakech Accords were

Table 1.1 Climate change conferences 1992–2013

Year	Conference	Notable Agreements
1992	Rio Earth Summit	United Nations Framework Convention on Climate Change (UNFCCC)
1995	COP 1 Berlin	The Berlin Mandate
1996	COP 2 Geneva	
1997	COP 3 Kyoto	The Kyoto Protocol
1998	COP 4 Buenos Aires	
1999	COP 5 Bonn	
2000	COP 6 The Hague	
2001	COP 7 Marrakech	The Marrakech Accords
2002	COP 8 New Delhi	
2003	COP 9 Milan	
2004	COP 10 Buenos Aires	
2005	COP 11/ MOP 1 Montréal	
2006	COP 12/ MOP 2 Nairobi	
2007	COP 13/ MOP 3 Bali	The Bali Roadmap
2008	COP 14/ MOP 4 Poznan	
2009	COP 15/ MOP 5 Copenhagen	The Copenhagen Accord
2010	COP 16/ MOP 6 Cancun	
2011	COP 17/ MOP 7 Durban	
2012	COP 18/ MOP 8 Doha	
2013	COP 19/ MOP 9 Warsaw	

Note: COP: Conference of the Parties; (MOP): Meeting of the Parties.

signed. The period preceding these Accords had been particularly acrimonious. Negotiations collapsed at COP 6 in The Hague in November 2000 after a bitter dispute broke out between the Americans and the Europeans over flexibility mechanisms. They were further disrupted by the inauguration of Republican George W. Bush as President in January

2001, who shocked the international negotiations in March that year by repudiating the Kyoto Protocol. Despite this setback, which many believed would kill the Protocol, furious lobbying in the intervening years by the Europeans, among others, led to the ultimate decision by the Russian Government to ratify the Kyoto Protocol in September 2004 securing its entry into force at COP 11 in Montreal in 2005 (Yamin and Depledge, 2004: Ch 2; Depledge, 2005: Ch 3). This decision marked the end of the Kyoto phase of negotiations ten years after they had begun at COP 1 in Berlin.

The latest phase of negotiations, the post-Kyoto phase, has attempted to reach agreement on what should replace the Protocol, which expired in 2012. This phase has seen a shift in the dynamics from negotiations centred around the US and the EU to negotiations centred around developed and developing countries (Bodansky, 2010). This reflects the more prominent role that developing countries, such as Brazil, South Africa, India and China, now have on the international stage. In December 2007, COP 13 in Bali set out the so-called 'Bali roadmap', which envisaged a path toward a legally binding agreement in Copenhagen at COP 15 in 2009 (Ott et al., 2008; Clemencon, 2008). Despite the expectations that the negotiations in Copenhagen, which included the presence of more than 100 heads of state, would produce a breakthrough and seal a new legal instrument, the negotiations stumbled. In the end, several key states managed to put together the text for the Copenhagen Accord, which was accepted as a political document rather than a legal one (Bodansky, 2010). The negotiations that have followed in Cancun, Durban, Doha and Warsaw have so far failed to agree on what should replace the Kyoto Protocol. As a result, the post-Kyoto phase of negotiations have quickly become another round of prolonged international negotiations that look set to continue for the foreseeable future.

CASES AND METHOD

Case Selection

To understand what factors lead a state to change its negotiating position and the type of agreement it is willing to sign in a prolonged international negotiation, this book examines two cases – the US and the EU – during the Kyoto phase of the climate negotiations. There are several reasons for selecting these cases. First, the US and the EU both actively participated in the Kyoto phase of the international climate negotiations, and critically, their negotiating position and the type of agreement that they were willing to sign changed over time. If one is interested in how and why the behaviour of states change in prolonged international negotiations one requires cases where this occurs. In addition, because the US and the EU have traditionally been critical to the success or failure of international negotiations, these cases also enable practical insights to be drawn for policymakers. For example, in the 1980s international agreement on the Montreal Protocol to address the hole in the ozone layer was not possible without the US and EU being in agreement. It was the same for climate change in the 1990s and, for that matter, negotiations on trade, finance and intellectual property. That is not to say that other major nations such as Russia, China or Brazil were not important, they were. However, if the US and EU were not in agreement international negotiations almost always failed.

Second, in order to critically analyse the three theoretical approaches across multiple case studies, cases were selected for the purpose of replication not representativeness. In the design of a case study, cases are not selected at random to be representative of a population. Rather the logic is analogous to that used in experiments. Similar cases are selected so that the findings of one case can be tested against a second, third or fourth case in an attempt to replicate the findings. The explanatory power of the theoretical approaches therefore can be tested across multiple cases enabling more robust findings (Yin, 2009: 54–55; Eisenhardt, 1989: 536–7; Silverman, 2001: 250–52).

An important step in replication is to ensure that cases are chosen in terms of the theoretical approaches to be tested. Accordingly, both cases share some fundamental characteristics, which are theoretically important. First, the US and the EU both share western democratic political systems. Second, because they do, the same sets of state and non-state actors, such as environment and energy agencies, and environmental NGOs and business groups, operate through the same democratic processes. For example, actors are free to engage in discussions, choose their own strategies and form coalitions, be it at the domestic, international or transnational level. This is critical if each of the theoretical approaches is to be properly analysed. In other words, in both cases actors operate with the same opportunities and constraints of a democratic political system.

Third, during the Kyoto phase of the negotiations all of the above characteristics remained constant. That is, the US and the EU remained democratic political systems, neither case experienced a fundamental change in the nature of their democracies. Finally, throughout this period there were no major exogenous shocks that transformed the context in which one of the parties was negotiating. For example, both the US and the EU were exposed to the same changes in the climate science, neither case experienced an environmental shock, such as Chernobyl or Hurricane Katrina, or a major economic disruption, any of which would have changed the context in which one of the parties was negotiating. In other words, many of the possible exogenous factors that could explain changes in the behaviour of the US or the EU were not present.

Method of Inquiry

This book employs a small-*n* case study approach. This was considered the most effective method both for critically analysing the explanatory power of the theoretical approaches in relation to the research questions, and for developing new theoretical concepts. In order to address the central research questions and make causal inferences about the 'how and why' of state behaviour in prolonged international negotiations, process tracing was used to complement the comparison between cases. Process tracing is a method of within-case analysis, which examines in detail the observable implications of hypothesized causal mechanisms in each case (Bennett, 2007: 35–6). Because it does not rely on establishing causation through comparison, this approach avoids many of the criticisms associated with relying solely on comparative case study research (King et al., 1994).

Following the multiple case study design, data had to be collected on the behaviour of the US and the EU during the Kyoto phase of the international climate negotiations. However, it quickly became apparent that there is very little publicly available data on US and EU decision making in the course of the negotiations. It is not hard to see why. It is virtually impossible to view the domestic and international components of international negotiations directly, given their closed nature and alternative options, such as archival research and interviews with participants after the fact, are costly and laborious (Odell, 2006a). Further, most negotiators spend much of their working life criss-crossing the globe, are often bound by confidentiality, and rarely have the time or inclination to write about the negotiations in which they have been involved.

Accordingly, to properly examine the behaviour of the US and the EU, I relied on elite interviews with representatives from state and non-state actors who were intimately involved in the negotiations. First, guided by the emphasis of each of the theoretical approaches on different actors at different levels of analysis, a pilot round of semi-structured interviews was conducted in 2009 with the aim of identifying some of the key actors (King et al., 1994: 22–3). Using the UNFCCC list of participants for the international climate negotiations between 1995 and 2005, respondents were identified who had participated on behalf of the US, the EU, the UNFCCC Secretariat and associated non-state actors. In all, I conducted 23 formal pilot interviews as well as a handful of informal interviews with people on the fringe of the negotiations.

As a result of the pilot interviews, it was decided to narrow the subsequent data collection to three sets of negotiations in the Kyoto phase: the negotiations surrounding COP 1 in Berlin in 1995, COP 3 in Kyoto in 1997 and COP 6 in The Hague in 2000. This was done for three reasons. First, the pilot interviews revealed that the most intensive period of negotiations during the Kyoto phase occurred close to the signing of the Kyoto Protocol in 1997. Second, access to key decision makers from negotiations in the 1990s was much easier than access to negotiators in the current negotiations. In other words, it was much easier to arrange an interview with the former Environment Minister of Germany, than it was the current Environment Minister. Third, respondents were more candid discussing negotiations in the Kyoto phase than they were in the lead up to Copenhagen. In sum, this period generated a richer body of empirical data than could be expected from interviews on more recent negotiations, and because of the resource intensive nature of elite interviewing, a shorter period also enabled more detailed process tracing of US and EU behaviour.

Two further rounds of interviews were conducted in 2009 and 2010. This time, respondents were identified using the webs of influence approach employed by Braithwaite and Drahos in their study of *Global Business Regulation* (2000). In contrast to the atheoretical snowballing technique, this approach deduced from the first round of interviews that there were key webs of influence among state and non-state actors that are critical in understanding state behaviour. These strands were followed in the second and third round of interviews by asking respondents from the first round such questions as: 'You say that the US Department of Energy was critical in forming the US negotiating position at Kyoto, who would be the best person in the department to talk with about this?' In this way, respondents for the subsequent rounds of interviews were identified and access was facilitated by first round respondents permitting their names to be used to solicit interviews with these key people. In the US respondents included representatives from the White House, State Department, Department of Treasury, Department of Energy, Environmental Protection Agency, business groups and NGOs, among others. In the EU it included representatives from the relevant Directorate-Generals of the European Commission, a sample of representatives from the bureaucracies in the member states and a small number of business groups and NGOs. In all, I conducted 105 formal interviews (23 in the pilot round, 35 in the second round and 47 in the third round).

To ensure construct validity, the data from the interviews were analysed in three ways. First, the data were evaluated for consistency within each case. Data provided from one negotiator in one case were checked against the data provided by his or her colleagues on the same delegation. This is especially important in elite interviewing because of the risk that respondents may exaggerate the importance of their role in events (Berry, 2002; Delaney, 2007). Second, data from the interviews were compared against the histories of the negotiations and policy documents compiled from an earlier literature review. Finally, draft sections of the study were sent to key respondents to check for historical inaccuracies (Yin, 2009; Bennett, 2007). Once this validation process was complete a rich body of empirical data was available for analysis.

BOOK OVERVIEW

This book comprises three parts. Part One encompasses the introduction (this chapter) and the theoretical framework (Chapter 2). Drawing largely from the field of international relations, Chapter 2 reviews the evolution of theories of international negotiations. In doing so, it outlines three theoretical perspectives: a two-level intergovernmental perspective; a transnational perspective; and an international regime perspective. Each perspective represents a different lens through which to view the behaviour of the US and the EU.

Part Two shifts to the case studies and considers the behaviour of the US and the EU in the international climate negotiations. Across three sets of negotiations, Chapters 3 to 5 critically analyse each of the three theoretical approaches to assess their explanatory power and to identify what factors led the US and the EU to change their negotiating positions and the type of agreement they were willing to sign. Each chapter draws on primary documentation and interviews with policy-makers and stakeholders, including state actors, environmental NGOs, and business groups.

Chapter 3 begins by considering the type of agreement that both parties were prepared to sign at the first Conference of the Parties (COP 1) in Berlin in March–April 1995. The result was the Berlin Mandate, which recognized that the commitments under the UNFCCC were inadequate and that developed countries should take on stronger commitments. Yet the agreement did not mention emission targets and timetables and it excluded any new commitments for developing countries. Chapter

3 highlights the strong role a progressive coalition of environment agencies and environmental NGOs played in the decision of the US and the EU to accept this agreement. It also shows how these interests were supported by government leaders and were largely unopposed by other actors.

Chapter 4 continues the inquiry by asking what type of agreement the US and the EU were prepared to sign almost three years later at COP 3 in Kyoto in December 1997. Had their positions changed? And if so, what factors led to a change in position, and how and why were these decisions made? It shows that the US and EU positions did change and that their decision to sign the Kyoto Protocol, which set legally binding greenhouse gas commitments for developed countries and excluded any voluntary commitments for developing countries, was again widely informed by a progressive coalition of environmental interests.

Chapter 5 concludes Part Two by considering the final set of negotiations at COP 6 in The Hague in 2000. The Hague failed to produce an agreement with both the US and the EU much more cautious about the type of agreement they were willing to sign. In the US this reflected a regressive coalition of actors which came to dominate the US position, and in the EU it reflected a strong progressive coalition, which was concerned about preserving the environmental integrity of previous agreements, notably the Kyoto Protocol. As the last of the case studies, this chapter also highlights the observed variations in the behaviour of the US and EU across the three sets of negotiations.

Part Three concludes by drawing together the data presented in the preceding chapters. First, it provides a critical empirical analysis of the three theoretical perspectives outlined in Chapter 2, and second, it sketches an 'ideal type' framework for improving our understanding of the temporal dimension of prolonged international negotiations. In analysing the three perspectives, Chapter 6 argues that the two-level intergovernmental perspective most convincingly explains the negotiation position and the type of agreement both parties were willing to sign in each of the three negotiations, though it notes that a complete explanation requires the insights of the transnational approach and the international regime approach as well. However, returning to the central research questions, Chapter 6 highlights that the existing theoretical frameworks do not sufficiently capture the factors behind the changes in the behaviour of the US and EU.

Accordingly, Chapter 7 takes the more difficult step of combining our existing theoretical knowledge with the findings of the empirical analysis in order to consider prolonged international negotiations. Honing in on the central research questions, it identifies a general set of internal and external factors that explain how and why the negotiating positions and the type of agreement the US and the EU were prepared to sign changed. These are then used as a basis to sketch an ideal type framework in which negotiations can be usefully conceived of as an immature or mature game, where strategic opportunities exist for networked actors to constructively influence state behaviour. Eight strategies are suggested that traditionally weak actors, such as environmental NGOs or environment departments, can employ to steer prolonged international negotiations toward their preferred outcome.

2. Histories and theories of international negotiations

In 2008, I attended my first international climate negotiation, in the old German capital, Bonn. As negotiators whispered to one another and raised motions and objections, it was difficult to know what was going on, and why. Why did some countries raise the same point repeatedly, or others refuse to move on to the next issue, until some seemingly minor piece of negotiating text was removed?

I was not sure at the time. But what did interest me, was what motivated these countries and their representatives? Why did they sometimes seek to cooperate and at other times obstruct progress? Did this reflect their interests? And, how and why did their behaviour change from Bonn to Bangkok and on to the next negotiating round?

Fortunately, I am not the first to ask these questions and there is much we already know about negotiations. There is an abundance of theories that seek to explain how and why international negotiations matter in world politics. Most are connected to the field of international relations, which in the post-war era has sought to explain international outcomes by looking at the interaction of nation states. Realists first inquired into the motivations and strategies of states to explain international outcomes. Neo-realists asked about the structure of the international system and how this could constrain states. Later, liberals and then constructivists, observing greater patterns of cooperation between states in economic and security affairs, focussed on state preferences to understand why states cooperate.

Today many international negotiations involve multiple state and non-state actors that interact over many issues, in many fora, over many years. Increasingly international relations scholars use terms such as 'governance' and 'networks' to capture the complex interdependencies between different types of actors at different levels, be it the domestic, international or transnational level, which invariably shape international outcomes (Rosenau, 2000, 2007; Hafner-Burton et al., 2009; Lake, 2010).

This chapter reviews the evolution of theories of international negotiations. I begin by briefly charting the developments from the first generation of theories in the realist tradition, to the emergence of a second generation of theories in the liberal tradition. Most of this literature is concerned with explaining how and why states cooperate. I then outline three theoretical perspectives from the second generation of theories: a two-level intergovernmental perspective; a transnational perspective; and an international regime perspective. These perspectives are distinguished by the actors they emphasise – state and non-state – and the level of analysis – domestic, international and transnational. For an observer of negotiations, like I was in Bonn, each perspective is like another lens that reveals different, though not mutually exclusive, insights about state behaviour in international negotiations. I conclude this chapter by defining the concept of prolonged international negotiations based on this literature.

TWO GENERATIONS OF THEORIES

The first generation of international relations theories in the post-war era have been largely dominated by realism. Realists argue that the international system is characterized by anarchy (Bull, 1995). States could be viewed as unitary rational actors with stable and coherent preferences that seek to protect their territorial integrity. As states are the most powerful actors and there is no other actor that can regulate their interactions, anarchy follows. Consequently, realists argue that states are loath to cooperate. The free rider problem is a classic example where states will not cooperate because they can take advantage of the contributions of others while not contributing themselves (Wohlforth, 2008).

In the ensuing decades, 'neo-realists' have further developed the basic axioms of realism. Principally, neo-realists emphasise the structural constraints that states face in the international realm, which limit and constrain their choices (Waltz, 1979). Whereas realists stress the motivations and strategies of the state to explain international outcomes, neo-realists emphasise the pressures of anarchy where state behaviour reflects the incentives and choices on offer. The tendency within neo-realism to focus on these characteristics also resulted in the use of game theory as a central tool of analysis (Milner, 1992: 467). Game theory, like realism and neo-realism, is structural, strategic and rational, and international relations theorists, therefore, use game theory to focus on the state as an actor and consider the strategies that can rationally be used to further state interests (Jervis, 1988).

Since the 1980s a second generation of theoretical perspectives has developed in response to the approaches rooted in realism. The trend away from realist conceptions of state behaviour was forged in liberal accounts of international relations. Like realists, liberals view the state as an instrumentally rational actor, but rather than focussing on the power resources of states, liberals focus on state preferences. The central premise is that 'state preferences derived from the domestic and transnational social pressures critically influence state behaviour' (Moravcsik, 2008: 236). States act instrumentally in the international sphere, but domestically each state is a representative institution. The state acts as the 'transmission belt' by which the preferences of individuals and groups are translated into foreign policy.

The assumption within realism (and neo-realism) that states are the only significant actors and that we do not need to look inside the 'black box' of the state because states act in a unitary fashion led liberals, and later constructivists, to question the realist model of world politics as they observed greater levels of cooperation between states. First, liberal scholarship began to disaggregate the state to look at sub-units of government and non-state actors. Second, in inquiring into state preferences, liberal theory became much more sanguine about the possibility of cooperation between states than realists (Keohane, 1984).

The most fundamental departure from the realist conception of state behaviour has come from constructivists. Constructivism is a social theory that contends that knowledge and reality are socially constructed. It disputes the claim that rationality means only instrumental rationality and emphasises that calculating actors are sensitive and contingent to historical, social and normative contexts (Adler, 2002). While liberal theory began this trend by considering how state preferences are formed out of the domestic sphere, what distinguishes constructivism is that the influences on preference formation are social not material (Hurd, 2008). Scholars in this field also emphasise that state behaviour is affected by the logic of appropriateness (March and Olsen, 1998: 951). In other words, states make decisions based on what seems like the right thing to do.

THREE PERSPECTIVES ON INTERNATIONAL NEGOTIATION

These developments in international relations have greatly improved our understanding of inter-state relations and the prospects for international cooperation. They have also produced three distinct perspectives that help to illuminate state behaviour in international negotiations.

A Two-level Intergovernmental Perspective

Theorizing about international relations has been dominated by state centred approaches. In explicitly *intergovernmental* approaches international outcomes are determined by the interaction of unitary nation states and domestic politics is treated as a residual matter (Waltz, 1979).

However, over the last three decades a growing number of scholars have relaxed the assumption of the state as a unitary actor as they seek to peer into the 'black box' of domestic politics (Lake, 2008). The most influential framework is Robert Putnam's (1988) 'two-level game'. Putnam argued at the national level, domestic groups pressure their governments to adopt policies they support, while governments seek power by engineering coalitions among their national constituents. At the international level, governments want to satisfy domestic pressures, while limiting any negative consequences from foreign developments.

In this two-level game governments try to manipulate domestic and international politics simultaneously. At the international level (Level I) bargaining is between international negotiators who are attempting to reach a tentative agreement (also known in the literature as statesman or chiefs of government or COGs). The COGs are accompanied by their advisors and diplomats at the table. At the national level (Level II) separate discussions are taking place within the various constituent groups about whether to ratify the tentative agreement at Level I. The same COGs appear at this table, as well, but here they are accompanied by parliamentary figures, bureaucrats and interest group spokespeople.

For an agreement to be reached the tentative agreement at Level I must be 'ratified' by both parties at Level II, the domestic level. The term ratification is used generically to refer to any process that is required to endorse the agreement, be it a referendum or supportive public opinion. The key point is that at Level II each side has win-set: 'the set of all possible Level I agreements that would "win" – that is, gain the necessary majority among the constituents – when simply voted up or down' (Putnam, 1988: 437). Only when the tentative agreement reached at Level I falls within the Level II win-sets of both parties to the negotiations will an agreement be reached.

Accordingly, the larger the win-sets the greater the possibility of an agreement because win-sets are more likely to overlap. However, a large win-set can also be a bargaining disadvantage for an individual state. For example, if Player A's win-set is very large, Player B can push for more concessions because it knows that they will fall within the win-set of Player A. This paradox was first pointed out by Schelling who noted that 'the power of a negotiator often rests on a manifest inability to make concessions and meet demands' (1960: 19).

In the two-level framework put forward by Putnam the lead negotiators monopolize the external representation of the state, but in contrast to the unitary actor model they are constrained by domestic political dynamics. In opening the black box of domestic politics, Putnam argued that three factors are critical for determining the success or failure of an agreement.

First, like scholars in the liberal tradition, Putnam was concerned with the preferences of domestic actors and the distribution of domestic coalitions. For example, in line with earlier studies on bureaucratic politics, the preferences of government agencies and bureaucratic coalitions will be an important influence on the policy-making process (Allison, 1971; Druckman and Mahoney, 1977; Krasner, 1972). So too will the concentration of domestic producer and consumer groups and their capacity to form winning or veto coalitions to restrict the COG (Moravcsik, 1993). As will the networks between these state and non-state actors. For example, energy producers may form a coalition with energy agencies to pressure the COG. In addition, the relative homogeneity of coalition interests and the type of strategies they employ will also affect their capacity to shape the domestic win-set. For example, if a coalition of domestic producers, such as energy producers, have homogeneous interests and high levels of organization they are likely to be a very effective winning or veto coalition.

The preferences of domestic actors and the distribution of domestic coalitions will also be affected by the politicization or domestic salience of the issue. This will depend on the actors and the issues. For environmental issues, for instance, it is not uncommon for an 'attention cycle' to take place in which the salience of the issue peaks and troughs (Downs, 1972). When the electorate is not attentive one can expect COGs, concerned for their re-election, to weigh more heavily the preferences of organized interest groups and vice versa when there is high issue attention (Harrison and Sundstrom, 2007; Trumbore, 1998).

Second, the position of the COGs will also be affected by the structure of domestic political institutions. For example, in proportional electoral systems, such as those in Europe, domestic groups focus on election campaigns because the ratification of international agreements by the parliament is almost a formality. In contrast, in majoritarian systems, such as those in the US, domestic groups devote their attention to pressure points of ratification like the US Senate where a two-thirds majority is required. Several empirical studies of environmental issues like climate change have compared the proportional electoral systems of the EU to the majoritarian electoral systems of the US and Canada. These have concluded that proportional electoral systems more closely represent environmentally concerned voters' interest, often through the emergence of Green parties, as is the case in the EU whereas, in majoritarian electoral systems, such as those in the US, COGs traditionally appeal to the median voter, for which environmental issues are much less salient (Harrison and Sundstrom, 2007; Kroll and Shogren, 2008).

Third, despite the domestic political constraints of COGs, they can use their privileged position at both tables to manipulate their own win-set and that of other nations. While, the 'political complexities for the players in this two-level game are staggering', as Putnam pointed out, sometimes 'clever players will spot a move on one board that will trigger realignments on other boards, enabling them to achieve otherwise unattainable objectives' (Putnam, 1988: 434). For COGs there are various strategies that they can use to trigger realignments. A COG may intervene in domestic political debates to influence the preferences of domestic actors and the distribution of domestic coalitions. Or at the international level, for instance, side payments, like the promise of aid dollars or government payments to affected industrial sectors, can be employed to enlarge the opposing side's win-set. These strategies may increase the bargaining power of the COG or the probability of agreement when ratification is in doubt. Accordingly, in this approach the preferences of the COG and the strategies they choose are critical for understanding state behaviour in international negotiations and international outcomes.

In summary, the two-level model provides a 'theoretically distinctive' variant of the intergovernmental perspective because it relaxes the unitary actor model and stresses the interaction between the domestic and international levels (Moravcsik, 1993: 17). Although Putnam claimed that his work is no more than a metaphor, the two-level intergovernmental model can be used to derive hypotheses or predictions for international negotiations. These include:

- In international negotiations the COGs monopolize the external representation of the state;
- States largely respond to domestic pressures when forming their positions, especially the preferences of domestic actors and the distribution of domestic coalitions;
- The type of domestic political institutions, for example, majoritarian or proportional electoral systems, will affect states' positions and the likelihood of an agreement being ratified; and

- The preferences of COGs and the strategies they employ will affect the ultimate national positions leading into an international negotiation.

A Transnational Perspective

In the two-level intergovernmental perspective the state is no longer conceived as a unitary actor. State and non-state actors at the domestic level affect the ability of negotiators at the international level to reach an agreement that can be ratified. However, the COG monopolizes the external representation of the state because they aggregate and manipulate the preferences of domestic actors that remain contained at the domestic level. The transnational perspective takes a different view. Scholars in this tradition argue that 'transnational relations matter in world politics' and that state behaviour in international relations cannot be understood without taking account of the cross-boundary activities of sub-units of government and non-state actors (Risse-Kappen, 1995b: 280). In this view, even though the two-level game disaggregates the state, it is too narrow because it implies limited access to the international system, which 'no longer holds true in many issue areas' (Keck and Sikkink, 1998: 4). In other words, we must look inside and outside state borders.

The transnational literature developed in two complementary directions. The first stream considered the role of sub-units of government. In 1974, Robert Keohane and Joseph Nye used the term 'transgovernmental relations' to describe 'sets of direct interactions among sub-units of different governments that are not controlled or closely guided by the policies of the cabinets or the chief executives of those governments' (1974: 43). As part of the liberal critique of the realist assumption that states are unitary actors in the international system, Keohane and Nye sought to highlight the 'relatively autonomous activities of the lower levels of the bureaucracies', which act independent from higher authority in international politics (1974: 43). In contrast to the two-level approach therefore, COGs were not seen as having a monopoly on the external representation of the state.

In this view transgovernmental behaviour could take two cooperative forms: policy coordination and coalition building. First, policy coordination involves informal communication among low level bureaucrats, for example to manage air safety, to more regularized coordination, which can create attitudes and relationships that change policy, such as reductions in trade barriers. Second, transgovernmental coalition building 'takes place when sub-units build coalitions with like-minded agencies from other governments against elements of their own administrative structures' (Keohane and Nye, 1974: 44). Government agencies do this to improve their chances of success within government. For example, environment agencies in Mexico may seek the support of the US EPA to push environmental reform inside the Mexican government. Or, economic agencies in Europe may rely on US economic agencies to limit the economic consequences of an international agreement, with which they have less sway (McKeown, 2009).

Until the 1990s, this framework was largely ignored as international relations returned to a state centred model. However, in 1995 Risse-Kappen's volume *Bringing Transnational Relations Back In* renewed interests in this area by highlighting the role of transgovernmental networks in different issue areas. Since then, Ann-Marie Slaughter (2004) has gone so far as to explore 'a new world order' based on the idea of transgovernmental networks. She argues that the proliferation of these networks has arisen from the 'need for and capacity of different domestic government institutions to engage in activities beyond their borders, often with their foreign counterparts' (Slaughter, 2004: 12).

The second stream in the transnational literature focuses on the role of non-state actors across borders (Keohane and Nye, 1972). Scholars in this stream argue that the concept of transgovernmental networks is too narrow because it is limited to governmental actors. In this view, transnational relations, then, refers to 'regular interactions across national boundaries when at least one actor is a non-state agent or does not operate on behalf of a national government or an intergovernmental organization' (Risse-Kappen, 1995a). This literature is based on a more substantive critique of intergovernmental approaches in arguing that states have lost control over non-state actors who can organize and move across national borders, be they individuals, multinational corporations or advocacy networks (Lake, 2008). As a result, theorists in this field claim that the discipline is moving away from the study of 'international relations' and toward the study of 'global society' (Barnett and Sikkink, 2008; Lake, 2010).

In 1995, Risse-Kappen (1995a) set out the broad terrain for transnational relations research in asking: under what conditions do transnational actors succeed or fail in changing the policy outcomes of states? In their study, the success of transnational actors was dependent on the domestic structure of the state and the role of international regimes, which I will turn to in the next section.

One of the most influential attempts to analyse the effectiveness of transnational actors was Peter Haas' pioneering work on epistemic communities. That is, a 'network of professionals with recognized expertise and competences in a particular domain and an authoritative claim to policy-relevant knowledge within that domain or issue-area' (Haas, 1992: 3). Haas argued that the language of science is becoming a worldview that penetrates politics everywhere and hence could affect how states' interests are defined. This is especially so in issue areas with high complexity and uncertainty, though he recognized that there must be demand for such knowledge from policymakers. The empirical inquiries of Haas and others showed that the involvement of epistemic communities can promote organizational learning by helping to create shared understandings in their specialized field and hence to improve state cooperation (Haas, 1989, 1992; Braithwaite and Drahos, 2000; Ruggie et al., 2005; Raustiala and Bridgeman, 2007).

Other scholars have focussed on 'transnational advocacy networks', which are formed around shared principled ideas, instead of scientific knowledge and expertise (True and Mintrom, 2001; Schroeder, 2008; Tarrow, 2005; Smith and Wiest, 2005). According to Keck and Sikkink (1998: 2) a transnational advocacy network 'includes those relevant actors working internationally on an issue, who are bound together by shared values, a common discourse, and dense exchanges of information and services'. They argue that these networks are created, for example, when domestic actors find their influence over a nation state is blocked. Because transnational advocacy networks are not powerful in a traditional sense they rely on persuasion or socialization. Keck and Sikkink claim that their influence derives from strategies of persuasion through the quick movement of information and the framing of particular problems, staging symbolic events, calling on powerful actors for leverage and holding states to account for international commitments (Keck and Sikkink, 1998).

The steady growth in research on transnational relations, and transnational advocacy networks, has been especially evident in the field of environmental governance and climate change (Abbott, 2011; Andonova et al., 2007; Betsill, 2008; Betsill and Bulkeley, 2004; Downie 2013a; Giorgetti, 1999; Gulbrandsen and Steinar, 2004; Newell, 2000; Schroeder, 2008; Tarrow, 2005). Much of the literature has sought to investigate the success of these networks in adopting the strategies that Keck and Sikkink identified in influencing state behaviour, from framing environmental problems to holding states to account for their international environmental commitments. For example, Betsill and Bulkely (2004) have highlighted the manner in which transnational networks act as a means through which issues are reframed and reiterated in the context of climate change governance.

In summary, a distinguishing feature of this perspective is that many of these transnational actors, both sub-units of government and non-state actors, are organized in network forms. Indeed, this perspective is very much characterized by a focus on 'governance' and 'networks' where multiple actors work in concert across state boundaries to affect state behaviour and influence domestic and international policy outcomes. In others words, its emphasis is on global governance, in contrast to much of the literature that characterizes the intergovernmental tradition. As such, it is no surprise that a quite different set of predictions can be derived from this perspective. These include:

- In international negotiations the COGs do not have a monopoly on the external representation of the state;
- Transnational networks that use framing, symbolic events, the enrolment of powerful actors and naming and shaming can affect state behaviour by influencing domestic and international policy outcomes;
- Sub-units of government, or bureaucracies, deal directly with each other to coordinate policy and build coalitions, which can lead to policy changes in other nations; and
- Epistemic communities promote organizational learning by helping to create shared understanding based on their specialized knowledge.

An International Regime Perspective

The international regime perspective, like the two-level and transnational perspective, has its origins in the liberal tradition of international relations. The original concern of this literature was to show that, in spite of what realists and neo-realists claim, regimes are a necessary part of world politics (Levy et al., 1995). Regimes refer to 'principles, norms, rules, and decision-making procedures around which actor expectations converge in a given issue-area' (Krasner, 1983: 1). In this book, the international climate regime refers to the UN legal regimes beginning with the UNFCCC and continuing with the evolution of the Kyoto Protocol. While realists have asserted that regimes are merely epiphenomena, after more than three decades of research on international regimes this is regarded by most as an unsatisfactory interpretation (Breitmeier et al., 2006: 5).

In order to demonstrate that regimes matter and can potentially impact international political interactions like international negotiations, scholars have emphasized regime formation. For example, research considered the conditions under which regimes rise and fall and why regimes form in response to some problems and not others (Breitmeier et al., 2006; Hasenclever et al., 2000; Levy et al., 1995). However, much of this literature has neglected how regimes evolve over time through a period of continuous negotiation.

A second pillar of the research agenda has turned to look at regime effectiveness (Young and Levy, 1999; Breitmeier et al., 2006). In particular, it focuses on assessing how exactly international regimes operate to influence behaviour and in the process solve problems? And, for example, how much variance in the behaviour of individual actors can be attributed to the operation of an international regime, compared to other factors? (Young and Levy, 1999).

Yet specifying the ways in which regimes make a difference to the behaviour of states presents methodological challenges. Without delving too deeply into these issues, which have been widely discussed elsewhere (see for example, Breitmeier et al., 2006) one way, and the way with most practical relevance, as Levy et al. (1995: 287) point out, is to pinpoint specific mechanisms by which regimes make a difference to behaviour. In other words, the behavioural pathways through which regimes produce results.

In assessing the effectiveness of international environmental regimes, Young and Levy (1999) suggest a series of utilitarian and non-utilitarian pathways. The utilitarian pathways suggest that, to the extent that actors are self-interested utility maximizers, regimes can influence actor behaviour by affecting the cost-benefit calculations of actors. For example, regimes can affect state behaviour when they reduce the incentives of defection from collective action problems by sanctioning cheaters (Young and Levy, 1999). In addition, Young (1999) concludes that environmental regimes can also affect the cost-benefit calculations of actors by manipulating intangible costs, such as a country's reputation on the international stage, in contrast to material costs.

While utilitarian pathways assume rational unitary actors, the non-utilitarian behavioural pathways relax the assumption of rational utility maximizing behaviour. One such pathway is regimes as bestowers of authority. This pathway suggests that those who view the rules and norms of a regime as legitimate or authoritative will regularly comply with their requirements without estimating the costs and benefits of such actions. Regimes therefore influence behaviour through a process of socialization. Driven by the logic of appropriateness, actors accept that behaving in accordance with specific rules and norms is the right thing to do. For example, government agencies whose task it is to implement a regime's requirements will come to accept this as part of their organizational mission.

Second, regimes can act as learning facilitators by initiating processes that give rise to social learning. This can be apparent in new ways of understanding particular problems, new ideas about how to address problems and the process of implementing these measures. For example, Checkel (2001) argues that the preferences, beliefs and desires of actors are shaped by participation in the decision-making process of regimes. Thus, an actor's choice is a function of persuasion and learning where social construction is less strategic than deliberative. In the context of the climate regime, others have drawn on the notion of regimes as learning facilitators to show how international regimes can act as drivers and triggers of policy diffusion (Oberthur and Tanzler, 2007).

Third, regimes can act as role definers. Regimes operating at the constitutive level can shape the identities of actors and in the process influence the way actors behave in the roles they are assigned. Young and Levy (1999) argue that actors regularly take on new roles under institutional arrangements, despite their basic identities being established prior to the regime. In this light, the emphasis is on how regimes redefine old roles, create new ones, and allocate them among participants to the regime. This could be especially important in international environmental regimes where the evolution of ideas about the environment precipitate the definition of new roles for actors.

Fourth, regimes can act as agents of internal realignments. In response to charges of state centrism this pathway relaxes the unitary actor assumption. In doing so, it overlaps with the two-level perspective and the transnational perspective because it shifts attention toward the impact regimes have on both state actors and non-state actors. Accordingly, this pathway suggests that regimes can affect behaviour by creating new constituencies or shifting the balance of power among groups vying for influence within states. For instance, regimes can be used by advocacy groups to name and shame states that consider violating international rules and norms (Newell, 2000). Or, for example, interest groups may be able to exploit international rules to further their own goals, even when they are not completely consistent with the original purpose of the regime (Cortell and Davis, 1996).

In looking at the internal dynamics of states and other collective entities, this pathway suggests that international regimes will mediate the impact of transnational networks of state and non-state actors. This is acknowledged by scholars in the transnational perspective (Smith and Wiest, 2005). For example, Risse-Kappen has argued that the more an issue area is regulated by international norms of cooperation, the more permeable state boundaries become for transnational activities. He states that 'highly regulated and cooperative structures of international governance tend to legitimize transnational activities and to increase their access to the national polities as well as their ability to form "winning coalitions" for policy change' (Risse-Kappen, 1995a: 6–7). Others have gone as far as to suggest that the access international regimes and agreements grant to networks may be as important as the content of the agreement itself (Hafner-Burton et al., 2009: 573).

In drawing predictions from the international regime perspective it is important to be mindful of the overlap with the first two theoretical perspectives. First, the interaction between the domestic and international levels in the two-level framework cannot be understood in the absence of international regimes. International regimes help to constitute the two-level game; the interaction between both levels would prove mighty difficult in the absence of some basic rules and norms which international regimes provide. Likewise, transnational networks of state and non-state actors must be conceived in the context of international regimes. Transnational networks almost rarely form without international regimes, and at the same time such networks contribute to the definition of international regimes themselves. In short, the following predictions must be considered in light of the conceptual interdependence between the three perspectives:

- International regimes matter in world politics because of their capacity to influence the behaviour of state and non-state actors and the networks in which they operate;
- Regimes influence actor behaviour by affecting the cost-benefit calculations of actors;
- Regimes influence actor behaviour by establishing certain rules and norms as legitimate or authoritative;
- Regimes act as learning facilitators by initiating processes that give rise to social learning, such as policy diffusion;
- Regimes shape the identities of actors and in the process influence the way actors behave in the roles they are assigned; and
- Regimes act as agents of internal realignments by creating new constituencies or shifting the balance of power among groups within and across state borders.

PROLONGED INTERNATIONAL NEGOTIATIONS

Before I turn to examine the behaviour of the US and the EU, it is necessary to be clear about what I mean by prolonged international negotiations. While much of the literature on negotiations, especially in the realist tradition, has been concerned with one-off negotiations, I am not the first to note that international negotiations are not static and nor are the problems they seek to resolve (Spector and Zartman, 2003; Daugberg and Swinbank, 2009; Enia, 2009).

In noting that negotiations evolve, scholars have been largely interested in identifying different phases in negotiations. Two sets of phases can be broadly identified; pre-agreement phases and post-agreement phases. Pre-agreement phases include those phases first identified by Zartman and Berman (1982). First, the diagnostic phase preludes when the parties formally sit down at the negotiating table. In essence, 'it is the span of time and activity in which the parties move from conflicting unilateral solutions for a mutual problem to a joint search for cooperative multilateral or joint solutions' (Zartman, 1989: 240). Second, is the formula phase, a period of 'intensive negotiations' in which parties commonly establish a formula that 'provides a substantive framework for agreement and a set of criteria for resolving details' (Zartman and Berman, 1982: 143). In the third phase, the detail phase, the parties send signals to each other, exchange points, arrange details, and attempt to bring the negotiations to an end using deadlines.

Post-agreement phases, on the other hand, refer to the negotiations that take place after an agreement has been reached over the terms and obligations of international treaties. Spector (2003) identifies three domestic and three international processes that make up the post-agreement phase. Domestically he points to: ratification negotiations, which are required to achieve formal acceptance of the agreement, rule-making negotiations, which are required to implement the agreement at the national level, and negotiations concerning monitoring, reporting and enforcement to provide feedback to the international regime. Internationally, he identifies three parallel processes: regime formation negotiations over the agreed-upon rules and procedures, regime governance negotiations concerning monitoring, reporting and verification, and finally, regime adjustment negotiations. Of course, each of these phases are not necessarily sequential and many overlap (Spector, 2003; Jonsson and Tallberg, 1998).

What then, do I mean by prolonged international negotiations? Prolonged international negotiations are defined here as substantive international negotiations over a legally binding instrument that continue for five or more years, which begin with bargaining over a tentative agreement and conclude with bargaining over the ratification of that instrument. Because I refer to substantive negotiations over a legal instrument, this definition can include the negotiation of consecutive non-binding agreements on the path towards a legal instrument, either in a bilateral or a multilateral negotiation. In addition, five or more years is considered a prolonged period because of the natural variations that occur over this period via political and economic cycles, such as a change in government or an economic downturn, which produce ripple effects.

Further, much like the literature on the two-level game, I assume a bargaining phase and a ratification phase, which provides the 'crucial theoretical link' between domestic and international politics (Moravcsik, 1993: 23). The bargaining phase refers to the negotiations that set out the framework for an agreement and proceed to flesh out the details of such an agreement for adoption whereas, the ratification phase refers to the negotiations that take place after an agreement has been reached but before it enters into force. This definition then does not include what Zartman referred to as the prenegotiation phase, which preludes when the parties formally sit down at the negotiating table. Nor does it include those negotiations that Spector described which take place after the ratification negotiations at the domestic level and the regime formation negotiations at the international level. In other words, substantive prolonged international negotiations do not include the ongoing monitoring, reporting, verification and enforcement processes, which proceed for years, and often decades, as the parties meet to review the implementation of a treaty.

To illustrate, the international climate change negotiations that took place between 1995 and 2005 can be considered a substantive round of prolonged negotiations. COP 1 in 1995 can be considered the start of the so-called 'Kyoto phase' of negotiations because it is here that parties began negotiating a tentative international treaty. After the Kyoto Protocol was agreed to at COP 3 in 1997, the ratification phase took place, which concluded with COP 11 in Montreal in 2005 after Russia ratified the Protocol and it entered into force. In other words, this was the period of the substantive negotiations and not those negotiations which followed to review the implementation of the treaty. In contrast, the international negotiations on the ozone layer could not be considered a prolonged international negotiation because of the speed with which the substantive negotiations concluded. In fact, the negotiations for the Montreal Protocol, which commenced in 1986 with bargaining over a tentative protocol, were concluded less than three years later when the Protocol entered into force in 1989 (Benedick, 1991).

CONCLUSION: A THEORETICAL PERSPECTIVE

The transition from first generation theories in the realist tradition to second generation theories in the liberal tradition has significantly improved our understanding of international negotiations. As scholars have observed the growing complexity of world politics and the interdependencies between different types of actors at different levels of analysis, new lines of inquiry have emerged. These developments have been part of broader movement away from the study of international relations and toward the study of global society.

In the two-level framework, scholars have sought to explicitly incorporate the preferences and behaviour of domestic actors into explanations of international negotiation outcomes. The transnational perspective has looked inside and outside state borders to, for example, consider how transnational networks of non-state actors influence domestic and international policy outcomes. And, studies in the international regime perspective have inquired into how regimes influence the behaviour of state and non-state actors and the networks in which they operate.

Together, these three perspectives, in so far as they emphasise different actors at different levels of analysis, capture the main factors that are commonly used to explain outcomes in international negotiations. Yet stepping back, it is not clear how these factors change over time to affect state behaviour. Too often these perspectives have been employed in a static way. In other words, they look only at the end game, at one specific negotiation outcome. For example, the two-level intergovernmental perspective is valuable for understanding how domestic politics and international relations interact in a one-off negotiation. But, if one is to inquire into how domestic political dynamics change to affect international outcomes in prolonged international negotiations, the two-level approach is limited. This is because it ignores the temporal dimension. In focussing on the end game it looks only at the role of actors at a point in time. The aim here is to highlight the temporal dimension of negotiations in order to analyse variations in the behaviour of actors over time.

In the following chapters, I will employ these three theoretical perspectives to examine the behaviour of the US and the EU in the international climate negotiations over the course of the 1990s. The aim will not be to determine which perspective is right and which is wrong, especially given the conceptual interdependence between the frameworks. Instead, it will be to better understand what factors explain how and why a state's negotiating position and the type of agreement it is prepared to sign changes in prolonged international negotiations.

PART II

The case studies

3. Toward Berlin 1993–1995: environmental interests and a tentative agreement

On 28th March 1995, Angela Merkel, the then German Environment Minister, opened the first Conference of the Parties (COP 1) to the UNFCCC. Invoking the spirit of the successful Rio Earth Summit three years earlier, she called on delegates from 170 countries to overcome the great political challenges posed by climate change (Earth Negotiations Bulletin, 1995: 2; UNFCCC, 1995a).

Most attention turned to the obligations of countries, or in the words of the UNFCCC, the 'adequacy of commitments'. The main challenge was to decide whether the commitments of developed countries were 'adequate' to meet the ultimate objective of the UNFCCC to prevent dangerous climate change (UNFCCC, 1992). However, aside from deciding whether this non-binding commitment was sufficient and what additional targets may be necessary, parties also had to address a slew of related issues including: what commitments developing countries should have; the means by which countries could reduce emissions; and financial and technology transfers to developing countries (Rowlands, 1995; Victor and Salt, 1994).

Although there was no shortage of contentious matters to engage with at COP 1, reviewing the adequacy of commitments was without doubt the most politically charged task. Under the UNFCCC, developed countries (the so-called Annex I countries)[1] had agreed that they would accept the principle of 'common but differentiated responsibilities'[2] and make the greater effort to reduce greenhouse gas emissions. However, many now felt that this non-binding commitment would not be enough and that negotiations should move towards a legally binding protocol with more specific commitments (Earth Negotiations Bulletin, 1995).[3] At this point tensions arose as developing countries (non-Annex I countries) demanded to be excluded from any new quantitative commitments.

A related issue concerned market-based mechanisms for reducing emissions, specifically so-called joint implementation. That is, the means whereby one country can fulfil its obligation by helping to reduce greenhouse gas emissions in another country. The idea is for an investor in a developed country, for example, to sponsor a project in a host developing country, which leads to emission reductions and the transfer of technology and know-how. While the EU was sceptical about the concept, most developed countries supported the initiative. Developing countries, on the other hand, opposed it, fearing that it would enable developed countries to avoid their responsibilities and shift the burden of responding to climate change onto developing countries. Indeed some developing country delegates saw it as a new form of neo-colonialism (Rowlands, 1995: 148).

After two weeks of negotiations, the parties reached an agreement on what came to be known as the 'Berlin Mandate'. The mandate explicitly recognized that the commitments under the UNFCCC were not adequate and that work should begin towards a protocol or other legal instrument to be completed in 1997. The priority was to set quantified limitation and reduction objectives for the period beyond 2000, and to reaffirm the principle of common but differentiated responsibilities – implying that developing countries would be exempt from any new commitments (UNFCCC, 1995c). This was generally viewed as a positive breakthrough (Earth Negotiations Bulletin, 1995). On the issue of joint implementation, the parties agreed to a pilot phase during which any country may choose to participate, though any actions undertaken could not be used to meet existing commitments (UNFCCC, 1995c). The Berlin Mandate proved to be a defining document, which set the parameters for the international climate negotiations up to Kyoto and beyond.

This chapter will begin by considering the type of agreement that the US and the EU were prepared to sign at COP 1 in Berlin in March-April 1995.[4] Subsequent chapters will analyse changes in what the US and the EU were prepared to sign as the negotiations unfolded toward Kyoto in 1997 and The Hague in 2000. Drawing on primary documentation and interviews with policy makers and stakeholders, I will analyse the role of state actors, environmental NGOs and business groups in determining the reasons why the US and the EU took the decisions they did.

In doing so, I will confront the central challenge of this book, that is, to explain how and why state behaviour varies in prolonged international negotiations. In the following chapters, I will employ the three perspectives – a two-level intergovernmental perspective, a transnational perspective and an international regime perspective – to assess their explanatory power and to identify what factors lead a state to change its negotiating position and the type of agreement it is willing to sign.

WHAT TYPE OF AGREEMENT WAS THE US AND THE EU PREPARED TO SIGN IN BERLIN, 1995?

Agreement at Berlin would not have been reached had the US and the EU not been prepared to sign the Berlin Mandate. So what had they been prepared to sign? As Table 3.1 shows, the US and EU had very different positions, especially around commitments for developed and developing countries, and the use of flexibility mechanisms.

Table 3.1 US and EU positions at COP 1

Issue	US position	EU position	Berlin Mandate
Adequacy of commitments	Existing commitments not adequate	Existing commitments not adequate	Existing commitments not adequate
Developed country commitments	Policies and measures *not* emissions targets and timetables	Policies and measures *and* targets and timetables	Quantified Emission Limitation and Reduction Objectives (QUELROS)
Developing country commitments	Strongly advocated that developing countries should have commitments	Open to no new commitments for developing countries	Developing countries excluded from any new commitments
Flexibility mechanisms (i.e. joint implementation)	Strongly advocated joint implementation	Sceptical about joint implementation	Pilot joint implementation mechanism adopted

Sources: UNFCCC, 1995b, 1995c; Earth Negotiations Bulletin, 1995; and interviews.

On commitments the EU had been determined to push ahead and strengthen the obligations of developed countries.[5] The EU argued that an agreement in Berlin should mention specific emission targets and timetables for developed countries. Yet the US delegation was reluctant to follow this path, especially if developing countries did not take on obligations. As the chief US negotiator stated: 'we are not going to commit ourselves to things we cannot do. We will be fortunate if we can keep this treaty alive' (Brown, 1995). In contrast, the EU indicated it could accommodate developing country demands for no quantitative commitments. In the end the US succumbed to international pressure and supported the EU position that developing countries would not take on new commitments. However, the US did succeed in ensuring that specific targets and timetables for emission reductions were not included in the text.

Second, in return for compromising on developing country commitments, the US was able to ensure that joint implementation was included in the mandate. The US pushed hard for the inclusion of such a mechanism at Berlin, arguing that it would improve flows of clean technologies between countries and provide a cost-effective way to reduce global emissions (Earth Negotiations Bulletin, 1995: 5). Despite scepticism among many EU member states that this would be used by the US to avoid domestic action, its inclusion helped to bring the US on board.

What factors influenced the US and EU in coming to their ultimate positions? And how, and why, were key decisions made? The decision to sign the agreement in Berlin had been informed by developments over the two years since the Rio Earth Summit in June 1992. COGs may have agreed to the Berlin Mandate, but did they monopolize the external representation of the state, as the two-level intergovernmental perspective would predict? Was their decision largely in response to domestic pressures, especially the distribution of domestic coalitions? While the transnational perspective would predict that transnational networks also affect state decision-making, it must be established that these networks existed, and if so, the role that they played. Further, it is important to discover if the international climate regime itself influenced US and EU behaviour. Are there identifiable behavioural pathways, which affected the actions of state and non-state actors? For example, did the climate regime affect the cost-benefit calculations of actors, or initiate processes that give rise to social learning? The next section will address these and related issues, considering first the case of the US, and then that of the EU.

THE CASE OF THE US

State Actors: The Clinton Administration Comes to Power

The agreement that the US was prepared to sign in Berlin was made by the US Government. The US Government is not a unitary actor, it is comprised of sub-units who each have a role in determining the US position (Downie, 2013b). As Table 3.2 shows, the principal government actors were those involved in the inter-agency discussions at the national level, namely the White House, the Environmental Protection Agency

(EPA), the Department of Energy (DoE), the 'economic agencies',[6] and those involved in the international negotiations, principally the State Department.

Table 3.2 US state actors 1993–1995

State actor	General position	Key figures
White House	Progressive	President, Bill Clinton Vice-President, Al Gore
Council of Environmental Quality (CEQ)	Progressive	Kathleen McGinty, Chair of CEQ
National Security Council (NSC)	Progressive	Eileen Claussen, Special Assistant to the President, Senior Director for Global Environmental Affairs at the NSC
EPA	Progressive	Carol Browner, EPA Administrator Dennis Tirpak, Director, Policy Office David Doniger, Director, Air Office
DoE	Cautious	Hazel O'Leary, Energy Secretary Richard Rosenzweig, Chief of Staff Derrick Forrister, Special Assistant to the Secretary Abraham Haspel, Deputy Assistant Secretary
'Economic Agencies'	Cautious to regressive	Lloyd Bentsen, Treasury Secretary Robert Rubin, Director of the National Economic Council (NEC) Laura Tyson, Chair of the Council of Economic Advisors (CEA)
State Department	Progressive	Timothy Wirth, Under Secretary for Global Affairs Rafe Pomerance, Deputy Assistant Secretary for Environment and Development David Harwood, Special Adviser, Office of the Under Secretary for Global Affairs

Sources: Interviews and primary documents cited in text.

In January 1993, President Clinton succeeded President Bush in the White House. The inauguration precipitated an inevitable shake-up within government agencies and in the structure of the domestic and international discussions.[7] Under the Bush administration the US had been reluctant to act and many powerful White House figures had worked to prevent US action on climate change, including the former White

House chief of staff John Sununu (Agrawala and Andresen, 1999). The make-up of the Clinton White House suggested that the US would, in future, take a much more progressive approach. The new Vice-President Al Gore had campaigned strongly on the issue both as a Senator and as a Vice-Presidential candidate and he brought with him two former advisors to head the Council on Environmental Quality (CEQ) and the EPA respectively. The new chief US negotiator was also Gore's former Senate colleague, Timothy Wirth, who had championed climate change in the Senate. In line with what the two-level perspective would anticipate, the preferences of these statesmen were to have a direct effect on the ultimate US position.

The new administration was determined to 'distinguish itself' from the Bush administration on climate change (US-27). Indeed, consistent with the expectations of the regime perspective, the climate regime seems to have manipulated the intangible costs of the White House. According to senior White House officials, part of the desire was to restore the US' international reputation on climate change. As a senator, Al Gore had worked behind the scenes to persuade the Bush administration that they were mishandling the Rio preparation and in danger of humiliating and isolating the US. Hence, when he came into office, as one official explained, 'there was a lot of thought that there were important relationships to rebuild' (US-31).

The act of distinguishing itself from the Bush administration on climate change took place almost immediately, though as the two-level approach would expect, much of the new administration's plans were constrained by domestic political pressures.

On the 21st of April 1993, President Clinton, in a speech celebrating Earth Day, announced that the US would voluntarily commit to reduce its emissions to 1990 levels by the year 2000 (Agrawala and Andresen, 1999: 461). This 'kicked off' the inter-agency discussion on climate change (US-21; US-26), which would ultimately result in President Clinton's Climate Change Action Plan, later that year (US Government, 1993). However, the details of the President's Plan were directly influenced by the budget preparations, which included a proposal for a British Thermal Units (BTU) tax.

At the heart of Clinton's campaign for the White House had been a promise to reduce the deficit and stimulate the economy. From the moment the election was over, a proposal for a BTU tax had been central to this commitment (US-21).[8] The proposal was designed to raise the price of energy, increasing revenue for a stimulus package while at the same time reducing greenhouse gas emissions. Specifically, the BTU tax would have taxed all fuel sources based on their heat content (measured by British Thermal Units). Coal would therefore be taxed the most, followed by petroleum and natural gas, which is the least energy intensive. Although the BTU tax was framed as a budget measure, it was also the main method by which the US would reduce its emissions (Agrawala and Andresen, 1999; Bryner, 2000).

The proposal was quickly killed by Congress (Bryner, 2000). In May 1993, it narrowly passed through the House of Representatives, but it was defeated in the Senate (Agrawala and Andresen, 1999: 462). The difficulty of getting climate-friendly legislation through the Congress was to become even harder with the so-called 'Gingrich revolution' in November 1994, which gave the Republicans control of both the House and the Senate for the first time in forty years (Peterson, 2004). In fact, within months the Republican leadership had begun challenging the regulatory authority of the EPA and weakening environmental laws (Bryner, 2000).

The defeat of the BTU tax 'changed everything' (US-26). First, it directly affected the President's Climate Change Action Plan because it meant taxes were off the table but the goal of reducing emissions to 1990 levels remained. Second, if the US was going to take a market-based approach to reducing emissions the only alternative to taxes was emissions trading. This was not necessarily an unpalatable option. The concept of emissions trading was already on the radar within government because of the success of the 1990 US Clean Air Act Amendments. As one former US official put it:

> In this context market-based mechanisms really mean the choice between two: to tax greenhouse gases or are you going to do emissions trading. With the demise of the BTU tax there was clearly no political will to use the tax mechanism and so by default you really only had cap and trade as an option. There was also a euphoria over emissions trading because of SO_2 trading in 1990 (US-27).

With the President committed to releasing a climate action plan (US-27), the White House, specifically the CEQ, instructed the inter-agency process to find a politically feasible way of meeting the stated emission target. Under Kathleen McGinty, a 'very young but totally trusted' former staffer to the Vice-President Al Gore, the CEQ took a very active role in coordinating the discussions (US-36; US-21). The task was not easy because the 'loss of the BTU debate had made everyone skittish' (US-5). Moreover in 1993, climate change was still low on the domestic political radar, notwithstanding attempts by the White House to raise public awareness about the issue (McGinty, 1993). As a result, there was not much desire within the government to put forward another major initiative like the BTU tax.

> The big contentious issue was the BTU tax. A cap and trade program was never really on the cards … or major regulatory approaches … so there was a decision pretty early on to do this in a programmatic way through incentives and a lot of voluntary programs … The centre piece was going to be the BTU tax, as a major stick, when that failed all we were left with was carrots (US-27).

Indeed, the Climate Change Action Plan, which was released in October, contained 47 'actions' most of which were voluntary. They covered energy, transport, and forests, among other sectors, with a strong focus on industry linkage (US Government, 1993). While the Plan claimed that it would enable the US to meet its greenhouse gas target 'with domestic emissions alone', it also outlined an approach for joint implementation internationally (US Government, 1993: 27). The Plan was provided to the Intergovernmental Negotiating Committee (INC) in September 1994 as the US Climate Action Report, which was to lay the basis for the position that the US would take to Berlin less than 12 months later.

Although the White House, especially the CEQ, had been prominently engaged in the inter-agency discussions, it was not alone in shaping the early climate change initiatives of the Clinton administration (US-31). Other government actors, specifically the EPA and the DoE also had a voice in government decision-making. The EPA was led by Carol Browner, another protégé of Al Gore, who had been appointed after the Democrats took control of the White House in 1992. With respect to climate change there were two key offices in the EPA: the Policy Office, which was led by Dennis Tirpak, and the Air Office, which was led by David Doniger.

The EPA 'were typically the agency arguing most strongly for environmental positions' (US-27). While the agency was always going to be engaged on climate change because of its legislated responsibilities on the environment, it was especially active in driving emissions trading. The strong support for a trading scheme came primarily from the Air Office under David Doniger, which in the words of one EPA official, was 'uber pro-trading' (US-37). The EPA had experimented with domestic emissions trading programs as early as 1974 (Fisher-Vanden, 2000: 173). It had also been intimately involved in the Montreal Protocol in 1987 (US-22) and after the success of the 1990 the Clean Air Act Amendments, which established the first significant market for emissions allowances in the world with the creation of the sulphur dioxide (SO_2) trading regime to address acid rain (Pizer, 2004), the EPA had become a strong advocate for flexibility mechanisms.

While most of the EPA's input was through the domestic inter-agency discussions, this was not its only channel of influence. Some EPA officials worked outside the national polity, in contrast to what the intergovernmental perspective would expect. According to interviews with EPA officials and staff at the OECD, 'there was this network of EPA people' (EU-41) such as Barry Nussbaum, Nancy Keith and Paul Stultman, who were championing emissions trading to other nations as early as 1991, through workshops like those held at the OECD in Paris (OECD, 1992a; 1992b). Whether or not this represented a well-developed transgovernmental network, as the transnational perspective would expect, sub-units of government were beginning to deal directly with each other. That being said, the aim of this network was to share information with the EU about emissions trading, not to change US state behaviour.

The DoE was also active. The Energy Secretary under the new administration was Hazel O'Leary. Richard Rosenzweig, her chief of staff looked after the domestic discussions and Derrick Forrister, Special Assistant to the Secretary, focussed on discussions at the international table (US-6; US-26). Although there would naturally be differences between the DoE and the EPA, as O'Leary joked during a White House press briefing in 1993, both agencies were working more closely together than they had in previous administrations (O'Leary, 1993).

Along with the EPA, the White House had instructed DoE to work on the President's Climate Change Action Plan (US-26). Yet after the defeat of the BTU tax, both the DoE and the EPA were forced to look for other avenues to reduce emissions. The DoE had played a strong hand in negotiating the President's Action Plan with industry, which largely comprised voluntary measures. In particular, DoE devoted significant time and effort to *Climate Challenges* (US-35), which referred to letters of intent from utilities representing 60 per cent of US electricity generation to either return their greenhouse gas emissions to 1990 levels by the year 2000, or limit emissions under strict performance measures (US Government, 1993: 22).

According to DoE officials there was 'high' engagement from 'the top tier' of the agency (US-35) and it would seem that the DoE had formed a preference on the two main issues for Berlin. First, on flexibility mechanisms, like the EPA, the DoE supported emissions trading. Second, it argued that the US should not commit to targets in Berlin in 1995 and that developing countries should take on some form of commitment (US-20; US-6). In the end the two issues were closely linked for the

DoE. The senior ranks of the agency would consider some commitments as long as they were linked to flexibility mechanisms, like emissions trading (US-6; US-35). As one put it, 'the tougher the targets the greater flexibility should be' (US-35).

The State Department also influenced the US position, and not only because of its role leading the US delegation. The State Department under Wirth had close links with the White House. As a senator, Wirth had worked alongside Al Gore in championing climate change. He also had close links with the NSC and Eileen Claussen, who had come across from the EPA to be Senior Director for Global Environmental Affairs at the NSC, and together they worked to realize the new administration's desire to restore its international reputation and 'signal to the international community that we want to play in this game' (US-35; US-25; US-9). On the one hand, this reflected the personal beliefs of Gore and Wirth, in the words of the two-level perspective, the preferences of the COG. On the other hand, it also suggests that the international climate regime had manipulated the intangible costs of these actors, namely the US's international reputation.

At the same time, as the two-level perspective would predict, the White House, as the COG, was able to manipulate domestic pressures by affecting the distribution of domestic coalitions, in this case inter-agency coalitions. The strong links with the White House helped the State Department, under Wirth's leadership, to steer the discussions and hence the US towards the most progressive position possible in line with the preference of the Vice President. As one former State Department official claimed:

> Tim always said that we were trying to help the Vice-President to obtain a really good climate policy; that was our job (US-25).

As a result, some in the State Department argued that the US could or should take on binding targets (US-25). In doing so, according to officials intimate with the discussions, the State Department was aligned with the NSC inside the White House and the EPA (US-25). Of course, this inevitably led to tensions with other agencies, as one DoE official reflected,

> The State Department may have acted like a referee on international discussions, but not domestically. I do not care about State's views domestically, as long as the White House knew what we're doing that was fine (US-35).

The interesting point to note is that in the lead up to Berlin the traditionally powerful 'economic agencies' do not seem to have played an important role. As one official remarked, 'I do not recall that they were particularly active. They were involved, but they certainly did not dominate the process in any way' (US-21). Beyond the BTU tax, the economic agencies did little more than monitor the climate change discussions and most of the senior ranks of the agencies had not developed a firm preference on the issues to be negotiated, other than to advocate that the US take a cautious approach.

> They were occasionally [engaged] and they were monitoring it but I think for them at that point it was not all the way real. … we hadn't entered the realm of serious binding commitments, so the economic agencies were not totally focused on it (US-25).

Environmental NGOs: 'Inside the Administration'?

The picture presented thus far suggests that the decision by the US to sign the Berlin Mandate was made by state actors alone. However, as each of the theoretical perspectives acknowledges the state is not a unitary actor and non-state actors, such as environmental NGOs and business groups, also affect state decision-making. The two-level perspective anticipates that while the COG monopolizes the external representation of the state, the preferences of non-state actors are accounted for through the domestic polity. In contrast, the transnational perspective predicts that the COG does not monopolize the external representation of the state because non-state actors work across state boundaries to influence domestic and international outcomes. The international regime approach views things differently. It would anticipate that the climate regime influences the behaviour of state and non-state actors and the networks in which they operate via specific behavioural pathways. I will first consider environmental NGOs, which refer to both advocacy organizations and think-tanks. The next section will turn to business groups.

As Table 3.3 shows there was a diverse community of environmental NGOs operating in the US during the Kyoto phase of the negotiations. And, as Agrawala and Andresen (1999) point out, the change in administration produced a distinct realignment in the relationship with environmental NGOs. From the outset the 'NGO community was inside the administration' (US-36) with many of the political appointments to the new administration coming from environmental organizations, which had been in fundamental disagreement with the previous Bush administration.

Table 3.3 Prominent environmental NGOs in the US 1992–2000

Environmental NGOs	Type	Primary Interests
Center for Clean Air Policy (CCAP)	Individual NGO	Research
Climate Action Network (CAN)	Umbrella organization	Advocacy
Environmental Defense Fund (EDF)	Individual NGO	Advocacy
Friends of the Earth	Individual NGO	Advocacy
Greenpeace	Individual NGO	Advocacy
Natural Resources Defense Council (NRDC)	Individual NGO	Advocacy
Ozone Action	Individual NGO	Advocacy
Pew Center on Climate Change	Individual NGO	Advocacy
Resources For the Future (RFF)	Individual NGO	Research
Sierra Club	Individual NGO	Advocacy
Union of Concerned Scientists	Umbrella organization	Advocacy
World Resources Institute	Individual NGO	Research
World Wide Fund for Nature (WWF)	Individual NGO	Advocacy

Sources: Interviews and primary documents cited in text.

As one senior US official quipped:

> There were no real NGOs in the Clinton years, none of the NGOs were home because they were all working in the government. We went out to try and find NGOs and you couldn't because they were all in the administration. In the Bush years the NGOs viewed themselves as the government in exile (US-14).

According to interviews with US officials, many of these people were key figures in the inter-agency discussions (US-37; US-35; US-5). For example, as the head of the US delegation Timothy Wirth worked closely with Rafe Pomerance, Deputy Assistant Secretary for Environment and Development, who had worked in the NGO sector both at the World Resources Institute and at Friends of the Earth until being appointed by the Clinton administration in 1993 (UNFCCC, 1995a). At the EPA, David Doniger had also come from the environment community having been at the NRDC since 1978. Eileen Claussen, who at the time was at the NSC in the White House, had picked Doniger to come across to her former employer (US-22).

In the years leading up to Berlin, these relationships ensured that there was a constant interaction between the administration and environment community, who as 'a key constituency of the Democrat Party' were 'constantly brought in for ideas' (US-9).[9] However, to have their voices heard on key policy decisions, non-state actors must ultimately network with specific government agencies (Newell, 2000: 134). For environmental NGOs, such as the EDF and the NRDC, this often meant the EPA, given their legislative responsibilities, whereas business groups were more closely connected to the DoE, as I will discuss below.

While other studies have shown that a key determinant of the degree of influence that NGOs are able to exert over government decisions is the closeness of their relationships to decision makers, the re-alignment in the relationship between environmental NGOs and the Clinton administration does not necessarily mean that they had influence (Newell, 2000: 131). Although it is difficult to directly assess their impact, as others have noted (Porter et al., 2000), the Berlin Mandate that the US signed contained some of the key elements that NGOs had been lobbying for. For example, the US decision to sign the Mandate, despite the fact that developing countries would be exempt from any new commitments, was precisely what prominent US NGOs had been lobbying for – including those with links to the administration, such as the NRDC. However, many still believed the US had not gone far enough, the Sierra Club described the US 'performance' at Berlin as a 'shocking abdication of leadership' (Atkinson, 1995).

As the international regime perspective would expect, the desire of the Clinton Administration to restore the US's international reputation on climate change provided NGOs with an opportunity to push the US toward a progressive position (Busby, 2010: 127). For example, a member of one US NGO argued that much of the consultation with environment groups was because 'they cared about their reputation and what NGOs were saying' (EU-30). Rather than having a direct influence on the text of the agreement, the influence of the environment community is perhaps better seen as a type of 'accountability politics' (Keck and Sikkink, 1998) where NGOs used their relationships in the Clinton administration, especially with key figures in the inter-agency discussions to push the US toward signing an agreement for the sake of its reputation.

The discussion so far indicates that NGOs were largely focussed on influencing the US position through the domestic polity allowing the COG to maintain its monopoly on the external representation of the state, as the two-level perspective would expect. However, environmental NGOs did establish transnational networks. For example, NGOs such as Greenpeace, EDF and WWF coordinated their participation in the international negotiations under the umbrella of the Climate Action Network (CAN). The most visible means by which CAN sought to influence proceedings was through the publication of ECO, a daily newsletter designed to, among other things, discredit arguments put forward by opponents of action on climate change (Betsill, 2008). Yet interviews with US officials indicate that their greatest influence was through their established links at the domestic level. Even at the international negotiations most US NGO officials focussed their attention on their national delegation as they sought to reinforce their domestic message. There is little evidence to support the proposition of the transnational perspective that these networks played a big role in US decision making. This is confirmed in other studies, which have documented the role of NGOs in this period (Betsill, 2008; Newell, 2000).

Business Groups: Are They Mobilized?

Business groups were not as engaged as environmental NGOs in the period leading up to Berlin, nor did they have the same close links to the Clinton administration. As Table 3.4 shows, many business groups during this period represented the coal and oil industries, often described as the 'fossil fuel lobbies', these included companies, such as Exxon, Mobil and Texaco and umbrella organizations, such as the GCC, API and the ICC (Newell, 2000: Ch 5; Levy, 2005). According to Newell (2000: 99–100), the agenda of these groups was characterized by an emphasis on the scientific uncertainties, the damage to the economy that climate change policies could inflict, and the growing emissions of developing countries such as China and India. At this early stage in the negotiations, other industries that were notionally supportive of climate change action, such as the insurance industry, had yet to properly engage with the negotiations (Leggett, 1999). One interesting feature is that the umbrella organizations often took a stronger stance than many of the individual companies who were members. A former staffer at the GCC explained why:

> The GCC was a focal point and it was a group that could take the heat for speaking up. … Very few companies had an international presence and they wanted the GCC to carry their water, but they did not want to damage their own international commercial operations (US-33).

Like NGOs, business groups were forced to network with specific government agencies. With responsibility for energy policy in the US, fossil fuel companies and energy utilities were naturally inclined to work more closely with the DoE (US-5). Groups like the Edison Electric Institute were more comfortable with Hazel O'Leary, the Energy Secretary, than with other members of the administration, in part, because she came from the industry (US-26).[10]

A DoE official characterized the relationship:

> Following the Rio summit these [energy] companies did want to engage. They knew us in the energy department and they wanted their views in the policy process, rather than having to deal with the EPA. … They were cooperative, it was not like they like doing this, but they were happy as they could be. We also dealt with the environment community but they had better relationships with the EPA (US-35).

The evidence would suggest that the broad preferences of the energy and utility companies were taken on board by the DoE.[11] A case in point was the President's Climate Change Action Plan, described above, which produced a series of voluntary actions that DoE had successfully negotiated with utility companies. More importantly, it would seem the preferences of these groups factored into the department's stance during the inter-agency discussions on both the key issues of commitments and emissions trading.

According to a senior DoE official involved in the discussions,

> In the lead up to COP 1 the energy industry had an allergic reaction to [the phrase] 'targets and timetables'. They didn't like the aspirational targets at this very early stage … for domestic purposes the language meant a lot because the fossil fuel industry had defined targets and timetables as something they were opposed to (US-6).

And around the time of the BTU tax debate, this official went on,

> From the start the [Energy] Secretary's agenda was to move the US position to an emissions trading model, because some voices in industry were pro-emissions trading, but none of them were pro-tax. Utilities and some of the oil were pro-emissions trading (US-6).[12]

While it is difficult to assess the direct impact of business groups, their role in working to defeat the BTU tax was significant because it pushed the US toward emissions trading. As discussed, the BTU tax, which was supported by NGOs such as the NRDC (Nordhaus, 2008), met strong opposition from Republicans and industry groups. In fact, a domestic

Table 3.4 Prominent business groups in the US 1992–2000

Business Groups	Type	Primary Interests
American Electric Power	Corporation	Electricity
American Petroleum Institute (API)	Umbrella organization	Oil and gas
Business Council for a Sustainable Energy Future	Umbrella organization	Sustainable energy
The Climate Council	Umbrella organization	Oil and gas
Edison Electric Institute	Umbrella organization	Electricity
Exxon	Corporation	Oil and gas
Ford	Corporation	Automobiles
George Marshall Institute	Umbrella organization	Public policy
Global Climate Coalition (GCC)	Umbrella organization	Oil and gas
International Climate Change Partnership	Umbrella organization	Sustainable business
Mobil	Corporation	Oil and gas
National Association of Manufacturers	Umbrella organization	Manufacturing
National Mining Association	Umbrella organization	Mining
Peabody Coal	Corporation	Coal
Reinsurance Association of America	Umbrella organization	Insurance
Shell	Corporation	Oil and gas
Texaco	Corporation	Oil and gas
US Chamber of Commerce	Umbrella organization	General business

Sources: Interviews and primary documents cited in text.

coalition that included the National Association of Manufacturers, the American Petroleum Institute (API) and the US Chamber of Commerce spent more than $2 million on television and newspaper advertizing in an effort to mobilize public opposition and target key senators (Hilzenrath, 1993; Sewell, 1996; Agrawala and Andresen, 1999: 470). In an attempt to quieten the opposition, Treasury Secretary Lloyd Bentsen negotiated exemptions with some parts of the fossil fuel industry, but they failed to woo enough senators (Hilzenrath, 1993).[13] The defeat was a bitter one for the administration, as one White House official recalled, 'I still have the bruises from the BTU' (US-31). The use of Congress to defeat climate legislation and influence the US negotiating position was a model that was to be used frequently by business groups, especially after 1994. It is also a model entirely consistent with the two-level perspective, which predicts that a state's negotiating position will be affected by the power and distribution of such domestic coalitions.

However, like environmental NGOs, business groups also networked across borders, and as the transnational perspective would expect, one strategy was to enrol powerful actors outside the domestic processes. A case in point was the Climate Council, a US lobby group run by Don Pearlman, a lawyer and former member of the Reagan administration (US-19), who was said to represent oil and gas interests in the US. Pearlman formed alliances with OPEC countries, principally Saudi Arabia, who had a similar interest in seeing the negotiations stall (Leggett, 1999; Newell, 2000; Gelbspan, 1997). The following quote, though long, best summarizes what many interviewees confirmed (US-19; US-26; US-14).

> One of the things that he [Pearlman] would do was to activate the Saudis to make them very troublesome. They had unbelievable relationships with these countries. Half the time they would write the talking points for the Saudis. They used every available resource and on the international stage they would use oil-producing countries – they would say 'hey do you want some advice' and many of these countries were trying to figure out what hotel they were staying in … the negotiations were incredibly complex and some of these delegations had two people on them. That made for a willing recipient for assistance, which Don Pearlman and people like him were only too happy to provide (US-25).

Yet according to interviews with members of these groups '70 per cent' of their time 'was spent on the domestic side' because it was here that they could influence Congress and the electorate (US-33; US-28). Even the Climate Council's international activities were designed to complement their domestic lobbying. As other groups, such as the Business Roundtable pointed out, there were basic difficulties in establishing effective transnational networks.

> … we are not averse to international cooperation but it is difficult enough to get a consensus among our own members let alone in international groups (US-8).

In short, while business groups may have made good use of transnational networks where possible, it was the distribution of domestic coalitions of non-state actors that had the greatest influence on US behaviour, as the two-level perspective would predict.

Why did the US Sign the Berlin Mandate?

The two-level perspective most convincingly explains the reasons for the US negotiating position and its decision to sign the Berlin Mandate. The power and preferences of a progressive domestic coalition, comprising government agencies and environmental NGOs, supported by a willing COG in the form of the White House, largely shaped US behaviour. However, that said, there is some evidence to support the propositions of the transnational and international regime approaches.

First, the US position that the existing commitments were inadequate was a direct expression of the Clinton administration's desire to 'distinguish itself' from the Bush administration and to restore the US's international reputation. The fact that the President and the Vice-President and the coordinating agencies within the White House, namely the CEQ and NSC, supported this goal meant that there was little inter-agency discussion on this point. Further, it was strongly supported by all of the environmental NGOs, many of which were closely tied to the new administration. Of course, the concern of this progressive coalition for the US's international reputation also indicates the influence the international climate regime had in manipulating the intangible costs of these actors, as the regime perspective would predict.

Second, on binding targets the US position reflected a trade-off or compromise between the different agencies. As senior officials argued, both the EPA and the State Department supported the strongest position possible (US-27). However, the DoE did not. The DoE's preference for a voluntary approach closely reflected the preference of their non-state networks in the energy industry. It had already worked closely with the industry on voluntary measures in the President's Climate Change Action Plan and it was not about to advocate binding targets in the face of what one DOE official described as industry's 'allergic reaction' to such a phrase.

Third, there appears to have been more agreement in the inter-agency discussions about commitments from developing countries and emissions trading. However, consistent with the two-level perspective, this largely reflected the distribution of domestic coalitions, notably the existence of a more regressive coalition of Republicans and business groups. After the Republican takeover of the House in November 1994, most agencies acknowledged that if the US was to sign an agreement it would be preferable, given the political situation, for developing countries to have commitments, as stated in the US submission to the UNFCCC. Similarly, after the defeat of the BTU tax in which business groups had played a key role, there was a general belief among state actors, including the EPA and the DoE, that taxes were dead and that emissions trading was the only alternative. Despite some opposition in the environment community, the EPA supported emissions trading given its successful experience with the sulphur dioxide regime. At the same time, the DoE and some utility companies, though not the oil and gas groups, were in favour because of the flexibility it would provide.

However, as the outcome at Berlin showed, the US was ultimately prepared to sign an agreement that went further than what the inter-agency discussions had agreed to. First, despite being opposed to targets, the US delegation agreed to what was referred to in the Mandate as 'quantified limitation and reduction objectives within specified time-frames'. Second, and most importantly given the views of the agencies at the domestic table, the US also agreed that there would be no new commitments for developing countries (UNFCCC, 1995c).

The main reason that the US signed such an agreement was because the White House, as the COG, and the State Department had reached a conclusion that they must get an agreement in Berlin. This was confirmed by several US officials intimate with the discussions (US-25; US-6; US-9).

> The main reason we ended up accepting the no developing countries commitment was because the US wanted an agreement and they didn't want a failure. We really wanted to do anything to get an agreement (US-6).

However, given the obvious objection that some state and non-state actors would have to the Berlin Mandate, what enabled Timothy Wirth to sign the agreement? Three factors would seem to have been important and each largely supports the propositions of the two-level perspective. The first factor was the preferences of the COG. In 1995 the State Department had a relatively free rein from the domestic discussions to pursue its chosen course of action. Boosted by its close links with the White House, in particular, the relationships Timothy Wirth had with Eileen Claussen at the NSC and with the Vice-President Al Gore, the State Department had the necessary freedom to make decisions over the head of objections from other actors. As one State Department official put it:

> The reason Claussen and Wirth were able to get their point of view across was because the Vice-President was the guy who bullet-proofed them from the rest of the departments (US-9).

Second, in the early 1990s, climate change was not a politically salient issue, which in turn affected the domestic political incentives of key players and therefore the power and preferences of domestic coalitions.[14] As Wirth lamented at Berlin, there was 'no constituency' pushing for new greenhouse gas limits, the 'issue is just not on the radar screen in the United States. It's a very frustrating position to be in' (Kinzer, 1995). And as a former White House official confirmed, despite the early efforts of the new administration, 'the issue was basically non-existent in the 1990s' (US-31). Although this meant there was little pressure for action, it also meant that those state and non-state actors who would likely oppose such an agreement had not yet mobilized. For example the Department of Treasury, the NEC and the CEA, had not become actively engaged in the inter-agency discussions. In addition, the realignment in the power of coalitions within the executive muted the voice of these traditional opponents of environmental interests who were more inclined to pull their punches. As a DoE official remarked, 'in a Democrat administration the EPA is normally more powerful and any DoE official knows that' (US-20). Further, aside from the influential intervention of business groups in the BTU tax debate, their lobbying remained steady and had not yet reached the intense level of activity that preceded Kyoto.

Third, and related, COP 1 was not proposing to negotiate any specific binding commitments that would have tangible implications in the short-term. Instead, the focus was on reviewing 'the adequacy of commitments' of the UNFCCC and setting out a path toward a more substantive agreement in future negotiations. This was the very reason that the economic agencies 'were not totally focussed' on the discussions (US-25). While the senior ranks of these agencies were conscious of the impact climate change could have on their interests, competing policy priorities coupled with this fact enabled the State Department to make concessions that may have been vehemently opposed had some of these actors been engaged. In other words, by 1995 the international climate regime had not triggered the type of domestic realignments that the international regime perspective would anticipate.

In summary, the evidence supports the propositions of the two-level approach. Although transnational networks, such as CAN, did exist, they appear to have had little, if any, effect on the US, in contrast to what the transnational perspective would anticipate. Finally, aside from the desire of the White House to restore its international reputation, there is no evidence to support the other propositions of the international regime perspective.

THE CASE OF THE EU

State Actors: Environmental Interests Unopposed?

In the EU the decision to accept the Berlin Mandate was made by the Council of Environment Ministers (also known as the Environment Council). In the international climate negotiations the Environment Council acts as the COG, because the European Commission (EC) does not have a mandate to negotiate on behalf of the EU, as it does for trade negotiations. Officially, it is the Chair of the Environment Council, which is the minister from the country that holds the EU presidency, who speaks on behalf of the EU. In 1995, the Environment Council represented a union of 15 nation-states, supported by the EC, and the Chair of the Environment Council, which was France.[15]

Table 3.5 shows the principal state actors involved in the inter-agency discussions in the EU, namely the environment agencies, the energy agencies and the economic agencies in the member states, and DG Environment, DG Energy and DG Economic and Financial Affairs in the Commission (Downie 2013c). This section will trace the role of these state actors to examine the influence each had on the agreement the EU was prepared to accept. In what follows I will focus on the UK and Germany because in the 1990s they were the two most powerful member states in the EU and, as many interviewees pointed out, the positions they took to a large extent influenced the ultimate EU position.

In the early 1990s the UK and Germany had begun to take a strong stance on climate change. In 1989, British Prime Minister Margaret Thatcher told the United Nations general assembly in 1989 that 'the problem of global climate change is one that affects us all' (Thatcher, 1989). Since that point the UK had begun to shake-off its reputation as the 'dirty man' of Europe. In 1992 it set a target to reduce carbon dioxide emissions to 1990 levels by 2000 (Collier, 1997). In June 1990, Germany, a country with long-held green credentials, adopted a goal of a 25 per cent reduction in carbon dioxide emissions by 2005 compared to 1987 levels (Hatch, 2007).

The stance of both countries was also made easier by the fact that greenhouse gas emissions were already on a downward trend. In the UK, the liberalization of electricity and gas markets in the early 1990s resulted in a fuel switch from coal to gas, which significantly reduced

Table 3.5 EU state actors 1993–1995

State actor	General position	Key figures
Council of Environment Ministers	Progressive	Michel Barnier (France), Chair of the Environment Council
United Kingdom		
Department of Environment	Progressive	John Gummer, Secretary of State for the Environment Derek Osborn, Director-General of Environmental Protection within the Department
Department of Trade and Industry (includes responsibility for energy)	Cautious to regressive	Secretary of State for Trade and Industry Michael Heseltine, 1992–1995 Ian Lang, 1995–1997
HM Treasury	Cautious to regressive	Kenneth Clarke, Chancellor of the Exchequer
Federal Republic of Germany		
Federal Ministry for the Environment, Nature Conservation and Nuclear Safety (BMU)	Progressive	Federal Minister for the Environment Klaus Topfer, 1987–1994 Angela Merkel, 1994–1998 Ansgar Vogel, Director General, BMU
Federal Ministry of Economics (BMWi) (includes responsibility for energy)	Cautious	Günter Rexrodt, Federal Minister for Economics
European Commission		
DG Environment, Nuclear Safety and Civil Protection (DG XI)	Progressive	Environment Commissioner Carlo Ripa di Meana, 1985–1992 Ioannis Paleokrassas, 1992–1995 Ritt Bjerregaard, 1995–1999 Director General, DG Environment Laurens Jan Brinkhotrst, 1986–1994 Marius Enthoven, 1994–1996 Jorgen Henningsen, the Director responsible for climate change

Sources: Interviews and primary documents cited in text.

British emissions (Collier, 1997). A similar reduction occurred in Germany with economic restructuring following reunification in 1990.[16] In addition, having offered to host COP 1 in Berlin, the German Government under Chancellor Kohl and Environment Minister Angela Merkel made it clear to the rest of the world that climate change and the success of the conference was a priority for the government (Kinzer, 1995).

The negotiations in Berlin in 1995 also coincided with the accession of Austria, Finland and Sweden to the EU. All three countries had 'very progressive climate change policies', which had popular domestic support (EU-22). Because the comparatively strong environmental standards of these countries had been an issue during the national referendums for EU membership, the new members were granted an exception to continue these policies in the hope that the EU would gradually follow suit (EU-22). As such, the enlargement of the EU helped to create a Nordic block, supported by Austria, which tipped the balance further in the direction of a progressive climate position.[17]

The strong stance of the member states was reflected in the broader EU position. Even before the Rio Earth Summit in 1992 the European Community had set a target of stabilizing emissions at 1990 levels by the year 2000, which was not surprising given the large contribution that the UK and Germany could make to reducing EU emissions (Jordan and Rayner, 2010: 61–2; Hatch, 2007). Then, in February 1993, the 5th Environmental Action Plan, which was agreed to by the member states and the Commission, set climate change as a priority of EU environmental policy (European Commission, 1993).

The environment agencies in the member states and their ministers in the Environment Council had been at the heart of the EU position. According to senior officials in environment agencies in the UK, Germany, and the Netherlands, among others, the environment ministers who sat on the Environment Council, which acted as the COG, such as John Gummer, Secretary of State for the Environment in the UK, strongly supported action on climate change and hence an agreement in Berlin. As the international regime perspective would expect, for some member states this was partly an issue of reputation.

> There was a very strong desire for the UK to be seen as strong on environmental issues, the UK had been seen as a dirty man of Europe and then there was a series of Conservative environment ministers who were keen on the environment starting with Chris Patten, and who wanted to be seen on the front foot on environmental issues (EU-45).

This naturally led to progressive positions. In fact, in line with the rhetoric of the EU's submissions to the UNFCCC (UNFCCC, 1995b), all environment officials interviewed in the UK, Germany, France and The Netherlands argued that their environment departments had pushed for strong commitments for developed countries and the exclusion of developing country commitments in the first period. The Federal Ministry for the Environment, Nature Conservation and Nuclear Safety (BMU) had been particularly strong on this point arguing throughout the early 1990s in inter-agency discussions that Germany should take a target of a 25 per cent cut in greenhouse gas emissions (Hatch, 2007: 46). Such positions were also encouraged by the progressive climate change policies of the new member states – Austria, Finland and Sweden (EU-22).

While the environment agencies did not have it all their own way in the discussions, the opposition from other agencies appears to have been mute. As the two-level perspective would anticipate, this influenced the distribution of bureaucratic coalitions in a similar fashion to that in the US. According to officials who participated in these discussions, climate change was simply not high on the agenda in the early 1990s for those agencies responsible for economic and energy policy. For example, British officials asserted that 'the Department of Treasury and the Department of Industry were relaxed' (EU-37). Another argued that this was because 'there were no public expenditure implications so the UK Treasury was okay with it' (EU-46). Likewise a French official stated that in the lead up to Berlin the Ministry of Finance was simply 'not very interested' (EU-40). Admittedly, in Germany the Federal Ministry of Economics (BMWi), which has responsibility for economic and energy policy, had opposed many of the domestic climate programs that the BMU had tried to put in place (Haigh, 1996: 164–6). Yet in the discussions about the position the German Government should take at the international level the BMU proposals were generally 'accepted in the end', even if they were not enthusiastically supported (EU-32).

I have so far focussed on the member states and not the Commission. This is because in the early 1990s the Commission played a limited role in determining what the EU was prepared to accept in Berlin. DG Environment was the agency responsible for climate change in the Commission. However, in the early 1990s the desire of the Environment Commissioner, Carlo Ripa di Meana, and DG Environment to 'demonstrate its international leadership' (EU-22) was limited by its capacity. As a senior DG Environment figure explained:

> When I arrived at the Commission in September 1994, there was very little prepared and people were looking to the member states and the Environment

> Council. The Commission did not play a big role, for example, there was no climate change unit in DG environment. And Environment was not a big Directorate General in the Commission in those years and it was less influential and was not seen as important as it is now.

The same appears to be true for the other Directorate-Generals. For example, according to an official in DG Energy, 'up to 1995 climate change wasn't even recognized as a problem within the Commission' (EU-51). At the time, the main focus of DG Energy was on an internal energy market within the EU. Similarly, in DG Economics and Finance (DG EcoFin) the senior ranks of the agency 'did not care about climate change' (EU-53). That being said, lower level officials within DG EcoFin and DG Research were looking at possible economic instruments to limit greenhouse gas emissions, yet in the years leading up to Berlin much of this work had not factored into the EU position at the international negotiations (Directorate-General for Economic and Financial Affairs, 1992a, 1992b; di Valdalbero, 2010) (EU-53; EU-60).

The Commission's proposal for a carbon tax was the exception. After the environment agencies in the member states had pushed for ambitious positions and targets, it was mainly left to the Commission to consider how this could be achieved EU-wide. Prior to the Rio Earth Summit, an inter-service group was established in the Commission to consider a carbon tax. Driven by the Environment Commissioner, Carlo Ripa di Meana, who vigorously advocated a carbon tax, it had the support of DG Environment and DG Energy. Importantly it was also supported, albeit conditionally, by Jacques Delors, the Commission President. This grouping was able to overcome opposition from DG Taxation and DG Industry, which had raised questions about the impact it would have on European competitiveness (Weale et al., 2000: 122). The final proposal, which was to be a key plank of the EU strategy to stabilize emissions at 1990 levels by the year 2000, was a tax shared equally between carbon content and energy content so as not to favour nuclear power.

While this progressive coalition may have succeeded inside the Commission, as the two-level perspective might expect, the success of the proposal was ultimately determined by the broader distribution of domestic coalitions. The Commission proposal met stiff opposition, especially from the UK and France (and from business groups as I will discuss below). For the UK, the issue was not a carbon tax per se, but rather that taxation was a sovereign matter for member states.[18] On the other hand, France supported a carbon tax, but vehemently opposed the compromised carbon *and* energy tax because it believed it would disadvantage its nuclear industry.[19] One interesting feature of the opposition was the seeming alliance among some economic agencies, which had been largely indifferent to the inter-agency discussions, but were eager to ensure that an EU-wide carbon tax was not introduced. For example, one senior UK official claimed that the UK Treasury Department 'tried to make sure that the Council of Economic and Finance Ministers got involved in the tax debate in the EU' because a 'finance minister is not going to sign a carbon tax even if an environment minister would' (EU-43). As a fiscal issue, support in the Council had to be unanimous for a tax to pass, and after four years of wrangling it was not. The tax was ultimately defeated in December 1994 (Jordan and Rayner, 2010: 60). The significance of this debate within the EU, and the divisions it created among government actors, was evident in the fact that six of the greener EU member states threatened to withhold their support for the EU's ratification of the UNFCCC unless the tax was adopted (Oberthur, 1999: 44).

While the tax debate was internal to the EU, much like the BTU tax debate in the US, it did help to frame the EU position in the lead up to COP 1. First, the defeat of the carbon tax proposal meant that the EU was not pushing other developed countries as hard to implement a carbon tax, as it might have been had the EU adopted the proposal. In fact, during the tax debate within the Commission, Jacques Delors, the Commission President, had insisted over the head of Carlo Ripa di Meana, that the tax be conditional on other OECD countries implementing a similar tax (*The Economist*, 1992). Second, and as a result, the EU was more open to the US concept of joint implementation. While some member states were strongly opposed to emissions trading, especially Germany and Austria (EU-52), 'there was no general EU position' on trading at the time (EU-22). In other words, the defeat of the carbon tax left the door open to emissions trading.

Environmental NGOs Networking with Environment Agencies

While state actors played a central role in determining what the EU was prepared to accept in Berlin, as was the case in the US, non-state actors can also play an important role. Was this the case in the EU? And if so, how were environmental NGOs and business groups able to affect the behaviour of the EU at Berlin? Turning first to environmental NGOs, Table 3.6 shows the principal environmental NGOs operating in the EU. In the lead up to Berlin there was not the realignment in the relationship between European governments and environmental NGOs, as there was in the US when President Clinton was inaugurated. In the UK and Germany the same governments had been in power since before the Rio

Earth Summit. Although both were from the conservative side of politics, as were several other European governments, environmental NGOs had a strong presence in Europe and many maintained good relationships with state actors.

Table 3.6 Prominent environmental NGOs in Europe 1992–2000

Environmental NGOs	Type	Primary Interest
EU Climate Action Network (EU CAN)	Umbrella organization	Advocacy
Foundation for International Environmental Law and Development (FIELD)	Individual NGO	Advocacy
Friends of the Earth	Individual NGO	Advocacy
Green Alliance	Individual NGO	Research
Greenpeace	Individual NGO	Advocacy
International Union for Conservation of Nature (IUCN)	Umbrella organization	Research
Royal Society for the Protection of Birds (RSPB)	Individual NGO	Advocacy
World Wide Fund for Nature (WWF)	Individual NGO	Advocacy
Wuppertal Institute for Climate, Environment and Energy	Individual NGO	Research

Sources: Interviews and primary documents cited in text.

However, as other studies have shown, what is crucial is the networks between state and non-state actors (Braithwaite and Drahos, 2000). In the EU most environmental NGOs networked with environment agencies, in large part because NGOs are generally seen as the 'natural hinterland' of these agencies (EU-49). As in the US, there was often a 'revolving door' between these actors (EU-38). For example, John Gummer's private secretary was Tom Burke, former head of Friends of the Earth (EU-37), and staff from European environmental NGOs spent time at DG Environment in the Commission (EU-49). Further, in some countries, such as the Netherlands, environment agencies provided financial support to environmental NGOs to subsidize their operating costs (Sewell, 2005: 104).

According to interviews with government officials the influence of non-state actors was strongest at the national level, especially where they worked in concert with state actors, as the two-level approach would expect. For example, NGOs helped to provide domestic political space for government actors who wanted to take progressive positions (EU-35). The British Secretary of State for the Environment, John Gummer, argued that it was 'important' for NGOs with large membership bases, such as the Royal Society for the Protection of Birds, to 'push the envelope' on climate change because it 'gave [him] the political room to move in cabinet' during discussions about the UK negotiating position (EU-48; EU-35).

What impact this has on the type of agreement that European governments like the UK were prepared to accept is difficult to ascertain. Many of the positions that the EU took to Berlin, such as not requiring developing country commitments to be on the agenda had been strongly advocated by groups such as CAN Europe, Greenpeace and WWF. That said, this may reflect a lack of engagement from actors opposed to such demands, rather than the strength of NGO influence. To be sure, when business actors did engage and initiated a vehement lobbying effort against the Commission's proposal for a carbon tax, as I will discuss below, they were able to overcome environmental NGOs like the Wuppertal Institute for Climate, Environment and Energy, which had lobbied in support of the tax (*The Economist*, 1992).

Outside the national level, European NGOs, along with their US counterparts, were members of transnational networks, such as CAN. Transnational networks of this kind can pressure state actors by generating politically usable information and moving it to where it has most impact (Keck and Sikkink, 1998: 16). Several EU negotiators acknowledged the significance of this role observing how NGOs act 'as a back channel helping you to know what other governments think' (EU-37). For example, a British official pointed out that they sometimes used CAN UK's links with CAN Japan to find out the Japanese position on certain issues (EU-56). An NGO delegate explained the role.

> We are able to pass sensitive information from one government to another through that shuttle diplomacy because one government will tell us information that they are not yet ready to tell the other government directly and then I can go say to Y this is where X stands (EU-30).

The influence that NGOs can have as an information conduit between state actors was evident in the final week of negotiations in Berlin with the establishment of the so-called 'Green Group'. With the negotiations floundering, NGOs worked to bridge the divide between progressive members of the G77 and the EU. After Germany and other EU member states had opened the door to an agreement by stating that they would not require developing country commitments to be on the agenda, NGOs toiled feverishly to bring the developing countries on board (Oberthur and Ott, 1999: 46). To a large extent, this was achieved through an intervention by WWF, among other NGOs, who drafted a one page document, which set out some basic areas of agreement between both blocs. The NGOs took it to the Indian negotiator 'who liked it' (EU-58) and then they were able to bring G77 to the table. As a result a 'Green Group' was formed between the EU and the G77, less the OPEC countries. The resulting Green Paper, which had its genesis in the original NGO document, became the basis of the Berlin Mandate (EU-58; EU-48). This account fits more closely with the transnational perspective than the two-level perspective, given the cross-boundary activities of these NGOs.[20] However, transnational environmental NGOs did not seem to influence what the EU was prepared to agree to, rather it was the distribution of the domestic networks of environmental NGOs and state actors, consistent with the two-level approach, which provides a better explanation for the influence for NGOs.

Business Groups: 'There was no Big Opposition'

In contrast to environmental NGOs in Europe, European business was less engaged, especially on the international stage where US business groups dominated. While EU business did belong to international associations such as the ICC, their focus was squarely on the domestic level – see Table 3.7. Almost all interviews with representatives from state and non-state actors in Europe claim that 'there was no big opposition' from European business to the international climate negotiations in the early 1990s (EU-35) and to the extent that they were a presence 'it was almost wholly driven by the oil industry' (EU-37). Yet even the oil industry was muted in its opposition compared to its counterparts in the US (Levy, 2005). As one British official put it: 'there was no Exxon in the UK' (EU-46).

As the two-level approach would anticipate, EU business sought to influence what the EU would be prepared to accept by lobbying through domestic processes. There is little evidence to suggest that business groups influenced EU behaviour through transnational activities, nor did they try to. A former European business representative explained why the 'most intense lobbying happens at the national level'.

Table 3.7 Prominent Business Groups in Europe 1992–2000

Business Groups	Type	Primary Interests
British Petroleum (BP)	Corporation	Oil and gas
Confederation of British Industry (CBI)	Umbrella organization	General business
European Business Council for a Sustainable Energy Future	Umbrella organization	Sustainable energy
European Chemical Industry Council (CEFIC)	Umbrella organization	Chemicals
European Confederation of Iron and Steel Industries (EUROFER)	Umbrella organization	Iron and steel
European Petroleum Industry Association (EUROPIA)	Umbrella organization	Oil
Imperial Chemical Industries (ICI)	Corporation	Chemicals
International Chamber of Commerce (ICC)	Umbrella organization	General business
Munich Reinsurance Company (Munich Re)	Corporation	Insurance
Shell	Corporation	Oil and gas
Union of the Electricity Industry (EURELECTRIC)	Umbrella organization	Electricity
Union of Industrial and Employers' Confederations of Europe (UNICE)	Umbrella organization	General business
World Business Council for Sustainable Development	Umbrella organization	Sustainable business
World Coal Institute (WCI)	Umbrella organization	Coal

Sources: Interviews and primary documents cited in text.

> Each national EU delegation has its own A4 folder on its national position. They go into the EU coordination meetings with their positions and they can't change that too much without Minister involvement. What this means for business is that if you want a position you need to get it into the A4 binder (EU-39).

The belief among business groups that 'most of the work has to be done at the national level before the negotiations', meant that they often had strong ties to state actors (Christophe Bourillon cited in Newell, 2000: 102). While environmental NGOs were closely tied to environment agencies, business groups generally had better relations with energy and industry agencies. For example, business representatives from the ICC, the WCI, and EUROFER all report having good contacts in industry and energy ministries (EU-39; EU-27; Newell, 2000: 99). As one put it, 'if we want to get something done we would go to the industry ministry' (EU-39). In the Commission however, it would seem that the lack of engagement from DGs, other than DG Environment, meant that business groups 'primarily engaged with DG Environment' and 'people like Jorgen Henningsen', who at the time was the director responsible for climate change in DG Environment (EU-39; EU-22). Nevertheless, other DGs, such as DG Energy, did from time to time come 'under pressure from energy producers' (EU-53).

These ties influenced the position of energy and industry agencies. For example, in Germany the position of the BMWi reflected the interest of the coal, steel and utilities industry (Beuermann and Jager, 1996: 189). In the Netherlands too, 'industry groups could rely on the Ministry of Economic Affairs to represent its interests in intra-governmental discussions' (Sewell, 2005: 116). And Cass argues that in the UK the British government has been 'a tenacious defender' of industry's interests in environment and climate change debates (Cass, 2007: 69). Interviews with UK government officials confirm this, and many pointed to the extensive consultation that state actors, such as the Department of Trade and Industry, had with business groups (EU-46). However, prior to Berlin most European business groups were not concerned about the international negotiations, in part, because they did not anticipate any tangible implementation requirements that would affect their interests. In other words, by 1995 the international climate regime had not triggered the type of domestic realignments that the international regime perspective would anticipate. As a result, there is scant evidence to suggest that these ties were used directly to influence what the EU was prepared to accept instead, they were likely employed for domestic purposes.

A case in point was the defeat of the Commission's proposal for an EU-wide carbon tax, in December 1994. The proposal 'spurred the massed ranks of Europe's industrialists to mount what is probably their most powerful offensive against an EC proposal' (*The Economist*, 1992). The Environment Commissioner, Carlo Ripa di Meana, described the lobbying effort as a 'violent assault' (cited in Sebenius, 1994: 294). Targeting the Commission, groups such as EUROPIA worked to water down DG Environment's original proposal. In the UK, the coal industry made sure that the House of Lords EU Committee, which criticized the proposal, was headed by Lord Ezra, a former chairman of British Coal (Newell, 2000: 104–5). In fact, largely because of business lobbying, energy intensive industries and major exporters were pre-emptively exempted from the tax even before the directive made it to the Council of Ministers (EU-24; Sebenius, 1994).

Why did the EU Agree to Sign the Berlin Mandate?

The interplay of state and non-state actors that resulted in the decision by the EU to accept the Berlin Mandate parallels that of the US and is best captured by the two-level perspective. State actors were central, as they were in the US, and the influence of non-state actors was largely incorporated in the domestic polity, as the debate over the carbon tax shows. However even more so than in the US, the negotiating position the EU took to Berlin largely reflected the preferences and power of a progressive domestic coalition of environment agencies and environmental NGOs. All environment officials interviewed stated that their environment departments had pushed for strong commitments for developed countries and the exclusion of developing country commitments, which was the ultimate EU position. This was also what environmental NGOs, such as EU CAN, Greenpeace and WWF had been calling for.

The fact that the interests of the environment coalition came to dominate the EU position is somewhat surprising. Traditionally environment agencies have a low standing in the inter-agency hierarchy (Collier and Lofstedt, 1997). For example, Beuermann and Jager argue (1996: 195–6) that the BMU was one of the 'less powerful ministries' in the early 1990s, in part because of its limited resources. A former Director General of DG Environment made the same point about the position of DG Environment in the Commission in 1994 (EU-22). Three factors are critical to explaining the dominance of the environmental coalition, and each largely supports the propositions of the two-level perspective.

First, the power of the progressive environmental coalition reflected the fact that many of the agencies that would normally be the natural opposition to the environment agencies, principally energy and economic agencies, were not mobilized. As discussed, climate change was not high on the agenda for these agencies, and 'because it was in the margins' as one environment official commented, 'you could get away with doing things that you could never get away with if everyone was focussed on it' (EU-37). This was because the negotiations were not threatening any tangible implementation requirements. As one former environment minister put it, 'I was not asking Treasury for any money' (EU-48). To be sure, when there was a direct threat to the interests of these agencies or their network of non-state actors – typically business groups – their opposition was strong and influential, as the defeat of the carbon tax highlighted. This also shows that at this stage in the negotiations there is little evidence to support the proposition of the regime perspective that the international climate regime would influence state behaviour by triggering domestic realignments.

Second, and consistent with the two-level perspective that the preferences of COGs will affect the ultimate position of a party, environmental interests were able to dominate because their position was supported by key leaders in the member states who wanted the EU to take a leadership position at COP 1 (Ringius, 1999b). This parallels what occurred in the case of the US. In the UK, the Secretary of State for the Environment, John Gummer, asserted that he was given a 'pretty free rein' by the Prime Minister John Major to push the climate change agenda. Likewise as the host of COP 1 the German Chancellor 'Helmet Kohl wanted to take the lead' (EU-32). The decision to lead was much easier to make in countries such as the UK and Germany where greenhouse gas emissions were on a downward trend and there was bipartizan support for action on climate change.

Finally, the preferences of the Environment Council were not simply to take a leadership position but to reach an agreement as well. This explains why the EU was prepared to accept an agreement that included a mechanism for joint implementation, which most member states remained opposed to or sceptical of. British and German negotiators stated that they were determined to 'keep the US inside the camp' (EU-45). While EU negotiators did attempt to prevent a joint implementation mechanism being included in an agreement, ultimately the desire by Angela Merkel, then German Environment Minister, and others, for COP 1 to succeed in reaching an agreement trumped any concerns about the use of flexibility mechanisms (EU-48; EU-32; EU-37).

While these factors are best captured by the two-level perspective, which predicts that domestic coalitions and the preferences of COGs will influence state behaviour, there is some evidence that transnational networks of state actors may have also played a role. Indeed, the issue of flexibility mechanisms could have been one area where transgovernmental networks were important. Recalling that some EPA officials had spent time with their European counterparts at OECD workshops discussing the operation of joint implementation mechanisms, this may have worked to allay European scepticism and hence facilitated the EU decision to concede on this issue.

CONCLUSION: A COALITION OF ENVIRONMENTAL INTERESTS PREVAIL

The decision by the US and the EU to agree to the Berlin Mandate is best captured by the two-level framework. To be sure, in both cases, the dominance of a progressive coalition of environment agencies and environmental NGOs, supported by willing COGs, who wanted an agreement, meant that they would collude as necessary to get an outcome in Berlin. In contrast to what the transnational perspective would predict, there was no wide-scale network of government actors or non-state actors outside of the national process, and the one or two transnational networks that did exist, such as the CAN, seem to have had little, if any, effect on the US or the EU. While there was no evidence to support the propositions of the regime perspective, such as the climate regime triggering domestic realignments, the one exception was the issue of reputation, which clearly figured into the thinking of the COGs as the climate regime manipulated the cost-benefit calculation of states.

The key actors in the coalition of environmental interests in the lead up to Berlin were the environment agencies and their associated network of environmental NGOs. The interests of these actors had a disproportionate influence on the negotiating position of both parties and on the type of agreement they were prepared to sign. In the US, the decision to accept commitments even though developing countries would be excluded from any new commitments was what key figures in the EPA and environmental NGOs had been calling for. Likewise in both the member states and the European Commission environment agencies led the push for the EU to take a lead position at COP 1. While they were not always successful, as the defeat of the carbon tax showed, in so far as the international negotiations were concerned, their positions were generally 'accepted in the end' (EU-32). In short, the story of Berlin is a story dominated by a coalition of environmental interests.

Three factors were critical to determining the power and distribution of domestic coalitions. First and foremost, unlike other agencies, environment agencies were engaged in the inter-agency discussions. The senior ranks in most environment agencies, often from the minister down were engaged, and the agencies had devoted sufficient attention to the issue in order to develop a preference on specific positions. As one British official observed, 'under John Gummer climate change rapidly became the number one issue in the Department of Environment' (EU-45). These state actors were often supported by environmental NGOs who were also active on the issue. Yet other powerful agencies and associated non-state actors, who have been the traditional counterweight in inter-agency discussions, especially the economic agencies, were largely indifferent. Aside from the debate over the BTU tax in the US and the carbon tax in the EU, these agencies and business groups had not actively participated in the discussions. As a result, environmental interests were largely unopposed.

Second, the negotiations in Berlin promised to provide the road map for negotiating a legally binding instrument. COP 1 itself was never going to negotiate a legally binding protocol. As a former Commission official put it: 'it was a wish convention not a real economic convention' (EU-49). This meant that most actors realized that the negotiations were about procedures and principles not about concrete commitments that could affect their interests. This helped to ensure that some state and non-state actors did not mobilize and hence had not yet made their voices heard. It also ensured that at this early stage in the negotiations the international climate regime did not significantly influence the behaviour of state and non-state actors as the regime perspective would anticipate.

Finally, the overriding objective among government leaders in both the US and the EU was for an agreement to be reached. In the US, the White House was determined to distinguish itself from the previous Bush administration and play a positive role in the international negotiations, which for the Vice-President, among others, meant reaching an agreement. In the EU too, government leaders, no less than German Chancellor Helmut Kohl, had publicly stated that they were determined to get an agreement. In other words, without rolling over, both parties were prepared to concede ground to make an agreement possible. In the next chapter, I will consider whether the same was true in Kyoto. What was the US and the EU prepared to sign three years later? What had changed? And, was the same coalition of environmental interests able to prevail?

NOTES

1. Annex I countries refer to those developed countries and countries with economies in transition to a market economy listed in Annex I of the UNFCCC.
2. Article 3.1 of the UNFCCC states that: 'The Parties should protect the climate system for the benefit of present and future generations of humankind, on the basis of equity and in accordance with their common but differentiated responsibilities and respective capabilities. Accordingly, the developed country Parties should take the lead in combating climate change and the adverse effects thereof' (UNFCCC, 1992).

3. In 1995, six months before COP 1 the Association of Small Island States (AOSIS), a negotiating bloc that included Tuvalu, Kiribati and Fiji, among others, submitted a proposal for a protocol to the Convention, which called for a 20 per cent reduction in greenhouse gas emissions by developed countries by 2005 (Oberthur and Ott, 1999: Ch. 4).
4. Although there was no formal 'signing' of the Berlin Mandate because it was a decision by the first Conference of the Parties to the UNFCCC, I use the phrase 'to sign' generically in the following chapters to confer acceptance of a decision or agreement.
5. German Chancellor, Helmut Kohl, set the tone at the outset of the conference when he said that a climate protocol must cover 'all greenhouse gases, lay down clear objectives and target dates for their reduction, and determine all measures to achieve this aim' (Kinzer, 1995).
6. 'Economic agencies' refers to the Department of Treasury, the National Economic Council (NEC) and the Council of Economic Advisors (CEA). While the latter two are technically located in the White House, all three overlapped in their work, personnel and position on climate change throughout the 1990s.
7. Up to the World Earth Summit in Rio in 1992, the EPA Administrator (the head of the EPA) led US delegations to international environmental negotiations (US-18).
8. During the transition period a non-official group was established to look at economic options for dealing with energy and environmental resource issues. The group was put together in December 1992 around the time of the two-day economic summit President Clinton hosted in Little Rock Arkansas. One of the early proposals for a BTU tax was produced by this group. The group was said to include Marc Chupka, Reed Hundt and Robert Sussman, among others (US-21).
9. According to one US official, the tone of many of these meetings said a lot about the nature of the relationships: 'What was really surprising in the Clinton years was that the NGOs that were there would 'bash and trash' the administration in the internal meetings. I was shocked by this, but the White House said we expect this because you always go after your friends. Then in the Bush years the meetings were much more cordial because the NGOs realised there was just no point' (US-14).
10. Prior to her appointment as Secretary for Energy, O'Leary had been an Executive Vice-President of the Northern States Power Company, an electricity and natural gas utility (US-35).
11. A member of the GCC described the industry's broad position in this way: 'Industry hated all the climate change measures, they wanted voluntary measures, but there was no need to do mandatory mechanisms they wanted nothing to do with this. If there were going to be mandatory mechanisms they would prefer market-based mechanisms' (US-28).
12. It is difficult to ascertain how much of a role the DoE had on the emissions trading discussions. The documentation and most of the respondents support the view that the EPA led the process on emissions trading. Yet some DoE officials claimed that 'the EPA hated this [emissions trading], they were a regulatory agency' (US-26), while another argued that 'it was us that took emissions trading to the White House not the EPA' (US-35). On balance it would seem that the DoE did engage in emissions trading discussions, though it is hard to conclude from the evidence that it was they, and not the EPA, who led on flexibility mechanisms.
13. While this is the public version of events regarding the defeat of the BTU, some respondents wondered aloud about the motivations of the Treasury Secretary. One claimed that Benson did not like the BTU proposal and the exemptions he cut with the coal industry were designed to 'open the floodgates' to exemptions for others, inevitably killing the proposal (US-26). While it is unclear if this is true, a member of the industry coalition, which opposed the BTU tax, argued that if it 'had not favored coal it would have gone through' (US-33).

14. For example, network news coverage of environmental issues dropped by 60 per cent from its peak in 1989 to 1993 (Carmody, 1995: 45).
15. While the EU does not fit neatly into a two-level frame, to the extent that it speaks with one voice at the negotiations, as represented by the Environment Council, it can be considered a two-level player. Hence the negotiations within the EU and the member states are considered as the domestic level and the negotiations in Berlin the international level. State actors in the EU therefore refer to both government agencies in the member states and Directorate-Generals (DG) in the EC. Transnational relations are considered to be relations with actors outside the EU, for example, between actors in the EU and the US, or in the EU and Japan; not between actors in France and Germany.

    The two-level game frame has been applied to the EU in different ways. First, some scholars have applied it to look at negotiations between national governments and their domestic constituencies on the one hand, and with the EU on the other (Moravcsik, 1993). Second, others have applied the two-level frame in a manner similar to that applied here, though specifically for trade where the EC has a mandate. The Commission acts as the chief negotiator on behalf of the member states at the domestic level and it negotiates with other states at the international level (Odell, 1993). For an overview of this discussion see (Pollack and Shaffer, 2001).
16. Between 1990 and 2000 Germany's greenhouse gas emissions reduced by 18 per cent. It is estimated that almost 50 per cent of the reduction was the result of economic restructuring during reunification (Schleich et al., 2001).
17. Norway had negotiated to accede to the EU, but its membership was defeated by the Norwegian electorate in 1994.
18. For example, John Gummer, the British Secretary for the Environment recalled a 'row' he had with another EU environment minister: 'He tried to get me to agree to an EU energy tax and I said listen, there are certain limits and one of them is no common taxation. There is no way you can get me to support this. I would veto it even if I support this or not the UK position is a matter of absolute belief that there should be no common taxation'. Another British official explained: 'you have to remember the context of the EU – in those days anything to do with the EU was a big political issue in the Major Government because most of the Conservatives were resistant to the EU agenda' (EU-45).
19. As one former French official observed: 'The problem for France during that period was that they actually had a surplus of electricity because of the nuclear development. So they wanted to export energy and so the Department of Industry and Treasury [sic] supported maximizing the consumption of energy and exporting it because of their surplus' (EU-40).
20. Also consistent with the transnational perspective is the case of Centre for International Environmental Law (CIEL), an environmental NGO, whose founding members, James Cameron and Philippe Sands, represented AOSIS countries at the negotiations in Berlin (EU-38).

4. From Berlin to Kyoto 1995–1997: rising opposition to environmental interests

Following the decision in Berlin in April 1995 to work toward a legally binding instrument, the parties to the UNFCCC met eight times.[1] The third Conference of the Parties (COP 3) in Kyoto, Japan was the culmination of this process. On the 1st of December 1997, COP 3 was officially opened. The President of the COP, Japanese Environment Minister Hiroshi Ohki, called on those present to establish a concrete international framework for protecting the global climate (Earth Negotiations Bulletin, 1997: 3). With over 10,000 participants representing 161 states and 278 observer organizations, the Kyoto negotiations were one of the largest international environmental negotiations ever held, and certainly one of the most anticipated (UNFCCC, 1997b).

In the three years to Kyoto, the regular rounds of negotiations had made some progress. The most important was the 'Geneva Declaration' at COP 2 in July 1996, in which the US for the first time agreed to accept a legally binding commitment (Earth Negotiations Bulletin, 1996). However, as the Kyoto negotiations began there remained a long list of substantive issues still to be resolved. The first set was around what was to be achieved, and by which countries. For developed countries, this centred on the level and design of the greenhouse gas commitment. Developed countries, including the US and the EU, had not agreed on the level of emission targets. In other words, how big the emission cuts would be. Nor had they agreed on the design of a target. In particular, which gases it would include and the baseline from which to measure the target. On the other hand, developing countries were adamant that in accordance with the Berlin Mandate they would not agree to any type of emissions target, despite US attempts to revisit the issue.

The second set of issues was around how any greenhouse gas commitments would be achieved. The JUSCCANZ[2] grouping of developed countries, which included Japan and the US, strongly supported the use of flexibility mechanisms, namely emissions trading and 'sinks', to help a state meet their target.[3] Developing countries, led by China and

India, vehemently opposed, while the EU remained concerned about the widespread use of flexibility mechanisms at the expense of domestic actions. The EU too had proposed its own type of flexibility mechanism, the so-called 'EU-bubble'. The bubble would allow the EU to take on a collective commitment under which some member states could increase their emissions while others could reduce theirs. Many states opposed this and saw it as hypocritical given the EU's stance on emissions trading. The EU also continued to argue for binding policies and measures. That is, a set of policies that developed countries would be mandated to adopt in order to reduce emissions, such as the phasing out of fossil fuel subsidies and energy efficiency programs. Again, parties disagreed on this issue, especially in so far as the policies would be binding.

After the first week of negotiations there was some convergence on these issues.[4] Yet as ministers began arriving on the 8th of December, with only three days to go, many substantive issues remained and some feared an agreement was out of reach. Attention turned to US Vice-President Al Gore, whose last minute decision to attend COP 3, it was hoped would help break the impasse (Helm, 1997). In his speech before the plenary, Gore famously instructed US negotiators 'to show increased negotiating flexibility if a comprehensive plan can be put into place' (cited in Oberthur and Ott, 1999: 86). As the speed and the intensity of negotiations increased, many of the critical negotiations around targets took place in a trilateral forum between the US, the EU and Japan. At 1am in the morning on the 11th December 1997, Chairman of the Committee of the Whole, Ambassador Estrada of Argentina, opened the final session of negotiations. With the clock ticking, Estrada and his gavel pushed delegates through the final hours, in what a Tanzanian delegate called 'negotiation by exhaustion' (Oberthur and Ott, 1999: 89).[5] While the negotiations threatened to collapse on several occasions, especially over emissions trading, after two weeks and a final negotiating session that ran through the night, the plenary adopted the Kyoto Protocol just before 3pm on the 11th December 1997, one day after negotiations had been scheduled to close (Oberthur and Ott, 1999: Ch 7).

Heralded as 'a major step forward' by government leaders (European Environment Council, 1997c), the Kyoto Protocol established a legally binding treaty in which developed countries listed in Annex B of the Protocol committed to reducing greenhouse gas emissions by an average of 5.2 per cent below their 1990 emission levels by 2012. Developing countries, on the other hand, were able to defeat a US proposal on voluntary commitments and so, in accordance with the Berlin Mandate, they did not have greenhouse gas commitments, though they could participate through the newly established Clean Development Mechanism, which was worked out between the US and Brazil. In return however, the G77 countries conceded to US and Japanese demands for an emissions trading mechanism to be included in the Protocol. Annex I countries would also be allowed to count some sink activities toward their emission targets. The rules around emissions trading and sinks were deferred for future negotiations (UNFCCC, 1997a).

The Kyoto Protocol was a remarkable success, which many believed was impossible at the time of Berlin. Indeed, it would not have been without the support and signatures of the US and the EU. But what type of agreement had both parties been prepared to sign almost three years after Berlin? Had their positions changed? And if so, what factors led to a change, and how and why were these decisions made?

In this chapter, I continue the inquiry into prolonged international negotiations by considering the three years following the negotiations in Berlin and culminating in COP 3 in Kyoto in December 1997. In order to make comparisons with the US and EU position before 1995, this chapter will follow the structure of the last. I will first examine the behaviour of the US, and then the EU, in the lead up to Kyoto. Throughout this chapter the two-level perspective, the transnational perspective and the international regime perspective will be employed to analyse the behaviour of the US and the EU up to, and at, Kyoto.

WHAT TYPE OF AGREEMENT WAS THE US AND THE EU PREPARED TO SIGN IN KYOTO, 1997?

As in Berlin, the role of the US and the EU was critical in determining the outcome in Kyoto. As Table 4.1 shows, while the positions of both parties had shifted since the Berlin Mandate, as negotiations moved towards the specifics of a legally binding instrument, there remained differences in their positions about what developed and developing countries should commit to and how any commitments would be achieved. The two sets of issues were directly linked.

As in Berlin, the EU had been determined to lead. In the words of British Prime Minister, Tony Blair, the EU had put its ‘cards on the table’ and it was time ‘for others to follow suit’ (cited in Ringius, 1999b: 12). The EU target was a 15 per cent emissions cut on 1990 levels. In response the US proposed stabilizing emissions at 1990 levels. However, the US made it clear that this offer was conditional on the inclusion of flexibility mechanisms, realistic targets and timetables, and ‘meaningful participation’ for developing countries. The problem for the US, as the

Table 4.1 US and EU positions at COP 3

Issue	US position	EU position	Kyoto Protocol
Developed country commitments Size of the target	Proposed stabilizing emissions at 1990 levels by 2012	A 15 per cent reduction compared to 1990 levels by 2010	A US target of 7 per cent and an EU target of 8 per cent reduction compared to 1990 levels for the period 2008–2012
Gases	Targets should cover a 'basket' of six gases	Targets should cover three gases (CO_2, CH_4, N_{20})	Targets to cover six gases
Budget period	Multi-year targets	3 to 5 year targets	First commitment (or budget) period is 2008–2012
Type of target (flat or differentiated)	Supports flat target	Supports flat target	Differentiated targets
Role of sinks	Sinks to be included with maximum flexibility and no limit on their use	Sinks to be included but a decision on the rules about flexibility and limits should be deferred	Sinks included with a decision on the rules deferred
Developing country commitments	Strongly advocates voluntary quantified targets (so-called Article 10)	Does not advocate targets for developing countries	No developing country targets
Flexibility mechanisms Emissions trading	Strongly advocates emissions trading with maximum flexibility	Emissions trading should supplement domestic action	Emissions trading included with rules on supplementarity deferred
Joint Implementation	Strongly supports joint implementation	Supports joint implementation	Joint implementation included
Clean Development Mechanism (CDM) Joint Action ('EU bubble')	Supports the CDM Opposed to EU bubble	Opposed to some parts of the CDM. Strongly advocates EU bubble	CDM included. EU bubble included
Policies and Measures (POMs)	Strongly opposed to specified mandatory POMs	Supports specified mandatory POMs	No specified mandatory POMs. The Protocol emphasises priority areas for policy action

Sources: European Environment Council, 1997a, 1997b, 1997c; Earth Negotiations Bulletin, 1997; Oberthur and Ott, 1999; UNFCCC, 1997a; and interviews. The UNFCCC Secretariat compiled responses from parties on issues related to sinks one week before Kyoto. See FCCC/AGBM/1997INF.2. 29 November 1997.

British Environment Minister Michael Meacher pointed out, was that 'Washington could not expect to win approval for its demands', 'unless it accepted a more ambitious target' (Clark, 1997). At the same time, the EU strongly advocated mandatory policies and measures for developed countries to reduce emissions, which the US 'firmly opposed' (cited in Oberthur and Ott, 1999: 105).

Clearly both sides were going to have to move if an agreement on a Protocol was to be reached; and in the end they did. The US agreed to a more ambitious target of a 7 per cent cut in emissions on 1990 levels, and the EU an 8 per cent cut. In return, the EU and the G77 abandoned their opposition to emissions trading. The US was able to ensure that emission targets included all six gases and that sinks could be used to offset emissions. Further, as one US negotiator put it, the EU 'finally threw in the towel' on binding policies and measures (US-27). The one area where the US was unsuccessful, which proved to be critical in the years that followed, was on developing country participation. Formerly known as Article 10, the US had proposed that developing countries take on voluntary commitments.[6] However, most developing countries fervently opposed the Article in principle. They argued that it betrays the Berlin Mandate, which excluded developing countries from any new commitments, and that now was not the time to address developing country commitments but to strengthen developed country commitments (Earth Negotiations Bulletin, 1997: 13).

The type of agreement that both the US and the EU were prepared to sign at Kyoto was different to that in Berlin. In the lead up to Berlin the decision to sign was essentially made by the COGs largely in response to the preferences of domestic coalitions, as the two-level approach would predict. Was this still the case? Or did transnational networks of government sub-units and/or non-state actors play a greater role, as the transnational approach would envisage? Further, by the time of Kyoto had the international climate regime also begun to influence US and EU behaviour? In short, what factors led the US and the EU to change their positions and agree to sign the Kyoto Protocol? It is to these questions that I now turn by first considering the case of the US.

THE CASE OF THE US

State Actors: 'The Whole Issue Got Elevated'

With President Clinton commencing a second term in the White House in January 1997, there was no drastic overhaul in the senior ranks of government agencies as there had been when Clinton was inaugurated in 1993, though this did not mean that these agencies played the same role in the internal discussions as they had three years earlier – see Table 4.2.

Table 4.2 US state actors 1996–1997

State actor	General position	Key figures
White House	Progressive	President, Bill Clinton Vice-President, Al Gore
Council of Environmental Quality (CEQ)	Progressive	Kathleen McGinty, Chair of CEQ
National Security Council (NSC)	Progressive	David Sandalow, Director for Global Environmental Affairs at the NSC
White House Climate Change Taskforce	Progressive	Derrick Forrister, Chairman
EPA	Progressive	Carol Browner, EPA Administrator Dennis Tirpak, Director, Policy Office David Doniger, Director, Air Office
DoE	Cautious	Federico Pena, Energy Secretary Richard Bradley, Director, Office of Global Environment Abraham Haspel, Associate Deputy Under Secretary
'Economic agencies'	Cautious to regressive	Robert Rubin, Secretary of Treasury Lawrence Summers, Deputy Secretary of Treasury Gene Sperling, Director of the NEC Janet Yellen, Chair of the CEA
State Department	Progressive	Under Secretary for Global Affairs Timothy Wirth, 1993–1997 Ambassador Stuart Eizenstat, 1997–1998 Eileen Claussen, Assistant Secretary of State for Oceans and International Environmental and Scientific Affairs Melinda Kimble, Acting Assistant Secretary of State for Oceans and International Environmental and Scientific Affairs

Sources: Interviews and primary documents cited in text.

The inter-agency discussions slowed after the Berlin Mandate was signed in April 1995 (US-37). However, the release of the IPCC Second Assessment Report in December 1995 and the subsequent 'Geneva Declaration' six months later at COP 2 precipitated a new set of inter-agency discussions. In July 1996 in Geneva, Timothy Wirth, the lead US negotiator, announced that the US for the first time would support a legally binding agreement (Earth Negotiations Bulletin, 1996). The shift in position came as a surprise to many, but seems to have been prompted by the release of the Second Assessment Report, which concluded that there was a discernible human influence on the global climate (IPCC, 1995). In fact, three months before the negotiations in Geneva, Secretary of State Warren Christopher had used the growing scientific certainty as a backdrop to announce a new 'green' foreign policy, calling climate change one of the most 'significant … challenges facing the world' (Harrison, 2000: 104). Interviews with US officials, including Robert Watson, who was the chief science advisor in the White House at the time, confirmed the impact the IPCC report had on the administration (US-36; US-9; US-7).

After the 'Geneva Declaration' the political terrain in the US shifted and with it the inter-agency discussions. According to White House officials, with climate change on the front page of major newspapers after the Geneva meeting,[7] 'there was a much more serious effort to think through climate change policy at a high level' (US-7; US-23). One official pointed out, that 'the whole issue got elevated to a deputy secretary level where the ticket to entry was that everyone had to have a briefing on the science' (US-7). As part of these changes, in February 1997, the White House Climate Change Taskforce was established under the CEQ and the NEC, with Derrick Forrister appointed as the Chairman. He, together with Katie McGinty, 'were calling the shots' (US-37). Part of its role was to help raise public awareness, and in the months that followed it coordinated a series of public relations events to that end.[8]

The release of the IPCC report, the subsequent Geneva declaration and the shift in the inter-agency discussions supports the propositions of the transnational perspective that: 1) epistemic communities promote organizational learning by helping to create shared understandings based on their specialized knowledge; and 2) international regimes trigger realignments, in this case helping to institutionalize the role of climate scientists in the policy-making process.

That said, the most significant change in the political terrain came with the Byrd-Hagel Resolution, which the US Senate passed 95–0 on the 25th July 1997, four months before the Kyoto negotiations were to commence. The non-binding resolution resolved that the US should 'not be a signatory to any protocol' unless it mandated specific commitments for developing countries, *and* would 'not result in serious harm' to the US economy.[9] The resolution had an immediate effect on the US negotiating position and much emphasis was made of the unanimous vote, especially by Republicans and business groups opposed to a protocol, as I will discuss later. One former administration official argued that the resolution 'was not a vote against a Kyoto treaty' (US-31), hence the decision by the White House to signal to Senate Democrats that they could vote for the bill, which helped to produce the 95–0 vote. However, the more likely reason, according to the majority of those interviewed, is that the White House was simply caught off guard. One former senate staffer bragged that Senator Byrd had 65 co-sponsors before the White House knew 'what was going on' (US-30). As a result, the administration 'tried to make it meaningless by voting unanimous' (US-22).[10] Although the White House and those senators who supported strong action were able to ensure that the resolution was accompanied with an endorsement of the IPCC's findings,[11] it was, as many interviewees conceded, 'a big mistake in hindsight' (US-22).

> At the time it was just thought that it was better to let it go through than to have a fight on it. But that was a big miscalculation as well because Byrd-Hagel took on this kind of iconic status, it became much more than a Sense of the Senate Resolution, it became much more than that (US-1).

Congress' new found attention for climate change highlighted the growing tension between the legislative branch and the position of the executive. As I will discuss later, the role that business groups played working in concert with members of Congress to pass the Byrd-Hagel resolution represented the emergence of a regressive veto coalition, which as the two-level perspective would predict, was to have a lasting impact on the US's role in the international negotiations.[12] For a start, it directly affected the emissions target. The White House was forced to design a target which reconciled the Berlin Mandate, which excluded any new commitments for developing countries, with the Byrd-Hagel Resolution, which stipulated the US should not sign any agreement, unless developing countries take on 'new' specific commitments. It is no surprise then, that in search of a politically workable position the White House announced various targets between 1993 and 1997, each slightly different from the next. In 1993, the President had called for greenhouse gas emission reductions by 2010, without elaborating on developing country participation. However, as pressure mounted in the final months before Kyoto, the administration began to articulate a new position. On the 6th October 1997, the President stated, in deference to the Byrd-Hagel resolution, that the US 'will ask for meaningful but equitable commitments from all nations [including developing countries] at Kyoto', in spite of the Berlin Mandate. Then two weeks later, on the 22nd of October, President Clinton finally announced that the US would stabilize greenhouse gas emissions at 1990 levels for the period 2008–2012 (cited in Cass, 2006: 130–131).

According to US interviews, each of these announcements, including the push towards binding targets, first announced at Geneva, reflected the preference of the White House at the time, or in the words of the two-level approach, the preference of the COG (US-31; US-37). However, they had not been made without inter-agency discussion, and while the White House had signalled what the target would be, other government agencies were left to work out how exactly these commitments would be met.

As in the early 1990s, the EPA was one of the most active voices in these discussions. Still under the leadership of Carol Browner, the EPA stood strongly in support of binding emission targets, but 'once the die was cast', as one official put it, 'that yes we are going to have binding targets, the question was how deep', and it was here that inter-agency tensions emerged (US-27; US 22). While interviewees did not mention specific numbers, there was a general belief within the EPA that emissions could be reduced without serious harm to the economy. This was in direct contrast to what the economic agencies argued (US-21).

The EPA's position derived from its positive experience with the 1992 Clean Air Act Amendments and the Montreal Protocol. First, as many officials argued, the EPA 'drew heavily' on the 'clean air experience' (US-1; US-5), to inform their climate change programs. Whereas in the lead up to Berlin SO_2 trading was not yet underway, by the time of Kyoto the cost of reducing emissions 'were more than reasonable' (US-27). The effect, as one EPA official claimed was that:

> People believed that because it was so successful and it clearly was, that this was going to be somehow easy to do for all greenhouse gas emissions, and that do not worry about the cost somehow miraculously emissions trading brings the cost down (US-27).

It would seem that a similar optimism surrounded the Montreal Protocol. Several officials claimed that the ozone regime had a 'large influence on people's thinking' and 'by the mid-1990s everyone was keen to plug the Montreal Protocol into climate change' (US-13; US-22). Another claimed that the regime provided 'optimism that we could find a technology solution. We thought the Montreal Protocol was proof that a Dupont would emerge in climate change' (US-7). As the regime perspective would expect, the international ozone regime initiated processes that provided learning opportunities for EPA officials, which they applied during the inter-agency climate discussions.

The State Department supported the EPA's position and there were close links between the senior ranks of both agencies and the White House. Indeed, together these actors formed the core of a progressive coalition within government, which as the two-level perspective would predict did influence the US position. After Berlin, Gore's former Senate colleague Timothy Wirth continued to lead the delegation, and Eileen Claussen, who had been at the EPA and then the White House, was appointed Assistant Secretary of State for Oceans and International Environmental and Scientific Affairs.[13] Along with Carol Browner, David Doniger and others in the EPA, this group was instrumental in the US decision to support a legally binding agreement at Geneva in 1996 (US-25; US-22; US-37; US-4). It also used the inter-agency discussions to argue that the US should adopt an emission reductions target, not simply stabilization of emissions, and State Department officials maintained their support for the use of flexibility mechanisms, like emissions trading, to meet such a target (US-9; US-22; US-25).

Shortly before Kyoto, changes in the senior ranks of the State Department coincided with intense inter-agency discussions as the US negotiating position was finalized. Within a matter of months Timothy Wirth and Eileen Claussen both left the State Department (US-37). While there is much conjecture about Wirth's departure,[14] the ultimate effect would seem to have been a softening in the State Department's stance. According to US officials, Stuart Eizenstat, who replaced Wirth, did not advocate climate change action as strongly, and he was more inclined to listen to the arguments of the economic agencies who were advocating a more cautious approach for the US (US-26; US-4; US-23).

According to a former State Department official, by late 1997 they 'had very limited [negotiating] instructions' (US-9). As a result, the State Department put ten questions to the White House around October to clarify their instructions especially on targets and developing country participation. After the Byrd-Hagel Resolution, the White House gave clear 'instructions that China, India and Brazil must have targets' though State Department officials were unclear 'whether they had to be binding or non-binding' (US-9). The White House also gave the State Department clearance to propose differentiated targets instead of flat targets during the negotiations, which they did. As to whether there was room to move on the size of the target, the same State Department official confessed that 'we didn't know what it was going to be we just thought we could do a little bit off a 1990 baseline' (US-9).

Whereas the EPA and the State Department advocated similar positions in the inter-agency discussions, DoE did not, and as one official pointed out, the 'EPA and DoE were often at each other's throats' (US-4). As in the early 1990s, DoE advocated a more cautious approach. Under the leadership of Frederico Pena, who had replaced Hazel O'Leary as Secretary of Energy in 1997, DoE supported a modest target with maximum flexibility. This meant opposing the EPA's preference for a strong reduction in emissions and pushing for a target to be designed to include 'all types of forestry', 'all six gases' and a stabilization period of 10 years (US-6; US-21).

While there were differences around the design of a target, there was more agreement on emissions trading. Like the EPA, DoE had become convinced about the benefits of emissions trading, as the sulphur dioxide regime began to show how emissions could be reduced at low cost. According to former energy officials, by late 1996 'we were all traders' (US-20) and 'everyone was drinking the [emissions trading] Cool-Aid' (US-21). Some of the energy officials had also been involved in the development of this system and hence had a 'real interest to see market-based mechanisms work' for climate change (US-35).

Consistent with the two-level approach, DoE also faced domestic political pressures, which were often channelled through Congress. One former senior DoE official recalled 'getting hauled to Capitol Hill and having to explain DoE positions in front of an aggressive committee chairman' (US-21). The increased pressure from the Republican controlled Congress was also evident with the release by DoE of a study by Argonne National Laboratory, which concluded that energy price increases resulting from emission reductions would lead to job losses as high as 200,000 (Fialka, 1997). A former DoE official recalled the event:

> The study sat there in the department for a while [we were] in the political hot seat because the Republicans found out about it and were trying to get it released. We got into a lot of trouble from the White House because it was not what the administration wanted to hear at that point. But because of the political pressure we had to release it (US-21).

After Timothy Wirth announced that the US would support a legally binding agreement in July 1996, economic agencies began to play a more active role. This was a major change since the negotiations in Berlin and was sparked by the recognition that binding emission targets would likely have tangible economic implications. In other words, by 1996 the international climate regime had begun to trigger the type of domestic realignments that the international regime perspective would anticipate. The first evidence of this was the change in the structure of the inter-agency discussions. In 1996, the NEC successfully lobbied the White House to co-chair the internal discussions with the CEQ.

Whereas Katie McGinty had previously led the discussions as head of the CEQ, in the lead up to Kyoto she shared the role with Gene Sperling at the NEC (US-31; US-25; US-27).

> … when we got closer to Kyoto the NEC said we need to co-chair this and so it was jointly done. This decision definitely came from the White House. But basically some of the big boys at the NEC said to the White House we need to start monitoring this deal (US-25).

According to interviewees, this changed the inter-agency discussions 'considerably' because the EPA, State and the DoE had to contend with growing input from economic agencies, who often took different positions (US-25; US-27). They became part of a more regressive coalition, which consistent with the two-level perspective altered the power and distribution of domestic coalitions. To be sure, unlike the EPA and the DoE, the economic agencies were reluctant to take measures to reduce emissions (US-35). One the one hand, this reflected differing positions on the costs of reducing emissions, and on the other, as some EPA and DOE officials claimed, it was because some of the senior ranks in Treasury 'were opposed to the overall issue of doing something on climate change' (US-35; US-31). For example, on the costs, economic officials criticized energy officials for being unduly optimistic. A case in point was the 'five-labs' study that the DoE released in 1997, which concluded that emissions could be reduced at a very low cost (Aldy, 2004). CEA officials labelled these the claims of 'techno-optimists' because of their optimism about how quickly new technologies would penetrate the economy (Frankel, 2001). In short, the economic agencies sought to push back on the preference of other government agencies that the US should take a strong progressive position at Kyoto.

However, it was not until the beginning of 1997, that the position of the economic agencies, or what one official described as the 'Treasury rearguard' was fully felt (US-4). At one of the earliest inter-agency meetings under the new co-chair arrangements, Lawrence Summers, then Deputy Secretary of Treasury, came in 'very heavily and very negatively' against the arguments put forward by other agencies that emissions could be reduced at a modest cost (US-25; US-23). The extent of the opposition from Treasury, in particular, was evident in the strong reaction of

Summers against statements by the CEA that depending on how they were designed, some greenhouse gas commitments could be achieved at a reasonable cost (US-23). It was then that 'something of an internal war began' (US-25; US-4).

The dispute between government agencies, and between economic agencies themselves such as Treasury and the CEA (US-23), largely focussed on the design of an emissions target. In other words, how much emissions should be cut, if at all. According to former US officials, Treasury Secretary Robert Rubin and Deputy Secretary Larry Summers, 'put their reputations on the line inside the White House arguing against US action' (US-23).[15] This would seem to have included arguing against a binding target and/or, that any target be as flexible as possible. For example, one official claimed that Treasury was pushing for a target over more than ten years (US-6). That the President went against their advice and announced an emissions target on the 22nd of October, as discussed, was seen as a 'big defeat', and the senior ranks of Treasury were 'shocked' that the White House would not decide in their favour (US-23). Although this may have been the case, there is little doubt that the active engagement of the economic agencies, which had not been there in the lead up to Berlin, did work to limit both the US negotiating position and what the US was ultimately prepared to sign.

Environmental NGOs Remain Close to the Clinton Administration

With President Clinton returned to the White House in January 1997, there was no realignment in the relationship between the administration and non-state actors. Key environmental NGOs, such as EDF, NRDC and the Sierra Club, maintained their close ties with the administration. And, as discussed in Chapter 3, many of the critical figures in the inter-agency discussions, like Timothy Wirth, Eileen Claussen and David Doniger, who had strong personal ties to environmental NGOs remained in the administration for much of the period leading up to Kyoto. As a result, environmental NGOs continued to be more closely tied to the EPA, and to a lesser extent the White House and the State Department, while business groups were more aligned with DoE (US-5; US-32).

However, these ties do not necessarily equate to influence. As other researchers have pointed out, determining whether such relationships changed US negotiating positions is fraught with difficulty (Newell, 2000: 124). That being said, there is some evidence to suggest that environmental NGOs had an impact on the Clinton administration. First, White House and State Department officials variously claimed that 'Green NGOs absolutely influenced us' (US-1; US-5; US-9). It seems likely that one area where this was the case was in disseminating the science. One NGO official claimed that environmental NGOs, under the coordination of US CAN, spent much of 1996 working to educate government officials and the public about the conclusions of the IPCC Second Assessment Report, which was released in December 1995 (EU-30). Given the ties between the senior ranks of the State Department and US NGOs, the fact that Timothy Wirth and the State Department relied heavily on the science to justify the declaration in Geneva in 1996 that the US would support binding commitments, could to some extent be put down to the success of NGO lobbying.

For scholars who study state and non-state actor networks, one of the most effective ways that NGOs can influence state behaviour is by infiltrating orthodox policy networks (Rhodes, 2007; Gough and Shackley, 2001). In the lead up to Kyoto, there is strong evidence to suggest that EDF, and to a lesser extent NRDC, did just that, in inter-agency deliberations on emissions trading (Betsill, 2008). According to interviews with US officials, EDF, which had played an integral role in developing emissions trading as part of the Clean Air Act Amendments in 1992, was critical in persuading the Clinton administration to take a similar approach on climate change (US-21; US-34). EDF staff, such as Daniel Dudek and Joseph Goffman, were 'in the room' with key EPA, DoE and White House officials when decisions were made about the US position on emissions trading, in large part because they were respected by government agencies as experts in the area (US-19). Former DoE officials also argued that the fact that EDF was supporting emissions trading made the senior ranks of the energy agency, including the Energy Secretary Hazel O'Leary, more comfortable because EDF was not seen as just another green lobby group linked to the EPA (US-26; US-35). While other high profile environmental NGOs, like the Sierra Club, opposed emissions trading (EU-30), that EDF was able to manoeuvre inside the policy network seems to have provided a unique opportunity to directly affect the type of agreement the US was prepared to sign at Kyoto. This evidence supports the proposition of the two-level approach that the preferences of non-state actors are incorporated through the domestic polity.

Nevertheless, while NGOs did engage in cross-border activities, in line with what the transnational perspective would anticipate, there is little evidence to suggest that this directly influenced US behaviour. For example, US NGOs, among others, continued to coordinate their participation under the CAN, as they had in Berlin in 1995, and organizations, such as Greenpeace International had a delegation of 45 representatives from across the globe at Kyoto (Leggett, 1999: 290–91). Further, US

CAN, as one former staffer pointed out, 'spent a lot of time trying to explain to developing countries that the Byrd-Hagel Resolution did not mean they had to take on targets' (EU-30). However, when asked about NGO influence, almost all respondents from state and non-state actors pointed to examples of NGOs working at the national level to secure their objectives, as the two-level framework would predict, and this is confirmed in other studies (Newell, 2000; Betsill, 2008).

Business Groups: 'Nothing Happens on the Hill by Accident'

Many of the same business groups that were present before Berlin continued to be involved in the US as the Kyoto negotiations approached. The most dominant continued to be the 'fossil fuel lobbies', such as Exxon and Mobil and their industry associations, the Global Climate Coalition (GCC) and the American Petroleum Institute (API). These groups, like 'most of US industry', were opposed to action on climate change, and many thought that 'global warming was not happening' (US-8). However, not all business groups took this position. For example, the International Climate Change Partnership, a coalition of industry representatives outside the energy sector, argued for a more constructive approach stating that climate change is 'an important matter with which government should be concerned' (Kevin Fay cited in Committee on Environment and Public Works United States Senate, 1997: 227–9). The Reinsurance Association of America went further arguing that climate change 'would bankrupt the industry' (White, 1998: 60). Nevertheless, it was those opposed to action on climate change that dominated, and the key difference since Berlin was that these non-state actors were now much more engaged.

The active engagement of business groups took place almost exclusively at the domestic level, as the two-level approach would expect. Like environmental NGOs, business actors lobbied the administration directly. For example, in July 1996, 119 CEOs and chairmen of companies such as Exxon, Ford, General Motors, Shell, Chevron and a host of coal companies, signed a letter calling on the President not to agree to any type of protocol (Leggett, 1999: 246). Many of these companies, especially in the utility sector, also engaged with the DoE, which as administration officials pointed out 'traditionally had very close links to industry' (US-5; US-26).

However, unlike environmental NGOs, business groups did not have strong ties to the Clinton administration, and as a result, they focussed their attention on the Republican-controlled Congress. In concert with Congress, business groups became central actors in a more regressive veto coalition. As the two-level approach would predict, the mobilization of this coalition shifted the power and distribution of domestic coalitions in the US. Nowhere was this more evident than through the Byrd-Hagel Resolution which, as noted, had a direct impact on the US negotiating position. According to interviews with former senate staffers and business representatives, the Resolution was instigated by many of the same fossil fuel companies and industry organizations that had defeated the BTU tax in 1993. As one respondent explained:

> The coal companies, the West Virginia coal companies … were telling us that we needed to do something on this issue. They wanted a Sense of the Senate Resolution that would put a stake through climate change and who best to turn to than Senator Byrd from their own state of West Virginia, who was a master of Senate procedures (US-30).

So he proved to be. With the support of Democrat Senator Byrd and Republican Senator Hagel, who the GCC had advocated as a co-sponsor, an extremely large and well-coordinated lobbying effort began (US-33). In the month prior to the Resolution, a meeting was held in Senator Byrd's office with the CEOs and Presidents of companies and industry groups in the coal, oil and chemical industries, including Peabody Coal, American Electric Power, and the GCC (US-30; US-5). According to interviews, each company was to focus their lobbying effort on two or three senators. A week later they reported back and the list of senators was flipped so that each company now had a different set of senators to target (US-30; US-33). As part of this effort, the National Mining Association sent around 40 CEOs to Capitol Hill in early June to speak with more than 50 senators about the Resolution. Labour Unions also took part (Sewell, 2005: 153). The lobbying also engaged local organizations, such as the chambers of commerce in each state, to further pressure senators (US-30; US-33). Business groups were aided by the mute response from environmental NGOs, who had been pre-occupied fighting attempts by the Republican Congress to weaken environment laws and did not foresee the significance of the lobbying effort and its ultimate result; the Byrd-Hagel Resolution (US-24). As one business representative proudly recalled 'nothing happens on the Hill by accident' (US-28).

At the same time, many of these actors helped fund advertizing campaigns to complement the message of the Byrd-Hagel Resolution that any treaty, which did not include all nations, posed a serious economic risk to the US (White, 1998). The largest was a series of television and newspaper advertizements, which ended with the tagline: 'if it's [the treaty] not global, it will not work'. Coordinated by the GCC and funded by member companies, principally Exxon and Mobil, it is estimated that business groups spent $12 million supporting the campaign (US-33). The campaign was extremely successful because, as one administration official conceded, 'it framed the debate' (US-1). As other researchers have pointed out, one of the most effective ways in which non-state actors can affect state behaviour is by strategically re-framing public debates to draw attention to their concerns (see for example, Braithwaite and Drahos, 2000; Joachim, 2003). Interestingly, while the transnational perspective points to the importance of framing, the evidence in this period indicates that it was employed most successfully at the domestic level not the transnational level.

Although business groups were most able to influence US behaviour at the national level, they continued to engage in transnational activities. For example, the GCC and the Climate Council took large delegations to the Kyoto negotiations. However as some of these representatives admitted, there was little strategic coordination with partner organizations in other countries such as the EU and Japan, as the transnational approach might expect, and their main aim was to influence the US Government (US-8; US-33). To some extent, this likely reflects the reality that US business actors can exert more power in the world system than those from other states because, as other researchers have observed, they can enrol the support of the most powerful state in the world (Braithwaite and Drahos, 2000: 490).

Why did the US Sign the Kyoto Protocol?

As in Berlin, the power and preference of a progressive domestic coalition supported by a willing COG, in the form of the White House, remained central to formulating the US negotiating position and determining what type of agreement the US would sign. However, by 1997 the dynamics of the domestic discussions had changed as a more regressive veto coalition emerged, which created divisions among state and non-state actors. First, the increasing engagement of the economic agencies led to splits within the executive branch of the US Government. Second, the legislative branch was increasingly at loggerheads with the position of the executive. The opposition from Congress was also supported by powerful business groups, who had been mute at the time of Berlin.

The two-level perspective best captures the significance of this distribution of domestic coalitions and the impact it had on the US negotiating position and its decision to sign the Kyoto Protocol. It also captures the role of the White House and the role its preferences played, informed by the science and its own domestic political incentives, in determining US behaviour. However, to completely understand why the US signed the Kyoto Protocol it is also necessary to consider the insights of the transnational perspective and the regime perspective, which both shed some light on the actions of the US.

Two sets of issues were critical to the US position. The first was the greenhouse gas commitments of developed and developing countries, and second, was how any such commitments would be met. First, the decision by the US to accept binding targets, and President Clinton's announcement before Kyoto that the US would stabilize emissions at 1990 levels by 2012, reflected the preference of the White House, the EPA, the State Department, and to a lesser extent the DoE, but not the senior ranks of the economic agencies. As several US officials pointed out, there were two groups within the administration; those that believed that addressing climate change would provide economic opportunities, and those that thought it would impose large costs on the US economy (US-31; US-27; US-4; US-25). The White House's decision to side with the former group, principally with the EPA and the State Department, was critical to the ultimate US position in support of binding targets.

However, unlike in Berlin, by 1997 a more regressive coalition had emerged. The Congress had actively engaged in the climate policy discussions. The effect was to force the US to formulate a negotiating position, which resolved to seek 'new' commitments for developing countries before the US would sign an agreement, despite what some government agencies such as the EPA preferred. With the active support of business groups, this was achieved with the Byrd-Hagel Resolution, which was passed unanimously by the US Senate, less than four months before Kyoto. As a result, the US negotiating position incorporated text on voluntary commitments for developing countries, so-called Article 10.

Second, as in the early 1990s, there was much less contention about how a target would be achieved. Once it became clear that the US would accept a binding target, there was unanimous support among government agencies that the aim should be to maximize flexibility. The instrument of choice was an emissions trading system. The EPA's positive experience with the Clean Air Act and the international ozone regime meant that they continued to support this approach, along with the DoE, as they had both done in the early 1990s. While the economic agencies may have disputed the costs of emissions trading, they were generally supportive of market-based mechanisms. In addition, environmental NGOs, notably EDF, were critical to building support for emissions trading within the administration by partaking in key policy networks.

Yet as it had done in Berlin, the Clinton administration signed an agreement that contravened the inter-agency conclusions. First, after Vice-President Gore instructed US negotiators to show greater 'flexibility', the US accepted an emissions target of 7 per cent below 1990 levels, instead of its original position to stabilize emissions. This was a significant shift in position and would require considerably more effort to meet than the EU target of an 8 per cent reduction on 1990 levels (Cooper, 2007).[16] Second, was the decision to accept a treaty that did not include new commitments for developing countries, despite the position of Congress. As one member of the negotiating team recalled, when it was removed from the text of the agreement, all 'hell broke loose because this was central to the US position' (US-26).

So, why did the Clinton administration decide to sign the Kyoto Protocol despite the opposition from a powerful regressive coalition? The short answer is they were determined to see an agreement reached. US officials argued that there was 'much pressure to come up with a deal' (US-37) and that the White House wanted an agreement (US-2). However, this only explains so much. The evidence suggests that three factors underpinned the desire for a deal, and while the two-level perspective best captures these factors, they cannot be completely understood without the insights of each of the three perspectives.

First, a deal at Kyoto formed a critical part of the White House's domestic political strategy. As the two-level approach predicts, the preferences of the COG will be informed by their desire to enhance their domestic political position. A former White House official explains:

> A lot of this was driven by the domestic constituencies, and their high expectations and we wanted to meet them. But it was easier to deliver on those expectations at the international level, where we had a free hand, than to deliver at the domestic level, where people are more constrained (US-1).

The desire to deliver a deal for domestic constituents, such as environmental NGOs, was also motivated by a second factor; scientific knowledge. The role of the IPCC, as an epistemic community in this process, is precisely what the transnational approach would anticipate. Since the Rio Earth Summit in 1992, a growing body of scientific evidence, culminating in the release of the IPCC Second Assessment Report in 1996, had concluded that human activity was the cause of climate change. With increasing interaction between scientists and White House officials, among others, this knowledge was beginning to permeate the policy elite by the time of Kyoto.[17] The Vice-President was at the forefront of this policy community and actively pushed the administration toward an agreement in Kyoto. According to US officials, it was Gore who said 'take the agreement' (US-2) despite opposition from the economic agencies. And, it was he, with the support of the President, who gave permission for Stuart Eizenstat, the lead negotiator, to negotiate a deeper emissions cut for the US.[18] In the end, the administration was able to marry the scientific consensus with their domestic political strategy because signing the Kyoto Protocol was very different to ratifying it, which was the Senate's job.

> The thinking of the administration was that if the Senate is going to reject Kyoto that is their business, but the Clinton administration wanted to be on the right side of history. … Climate change was something that they believed in, but it was also good politics because that is what the environmental NGOs wanted (US-34).

Finally, the US desire for a deal was also underpinned by concern about its reputation on the international stage. As it had in Berlin, the international climate regime had manipulated the cost-benefit calculations of the COG by raising the cost of reputation. In September 1997, three months before Kyoto, President Clinton announced that the US would not sign an international treaty to ban anti-personnel land mines, which had been negotiated in Oslo (Priest, 1997). The decision was strongly condemned by the international community, and according to former White House and State Department officials, it seems to have affected the thinking within the administration in the lead up to Kyoto. One senior negotiator recalled being given the instructions from Strobe Talbott, then Deputy Secretary of State 'don't let this be Oslo' (US-9). Another summed up the approach:

> Foreign policy was important. Our standing in the world was a big consideration as the Kyoto Protocol came in the wake of negotiations over a land mine treaty. We were alone in the world on that, so this was an important factor in our thinking: we don't want to be like that again – all alone (US-7).

THE CASE OF THE EU

State Actors: Environment Agencies and the 'True Believers'

Among the 15 member states, the UK and Germany maintained their position as the two most powerful states and both were resolute in their conviction that the EU should take a leadership role in the climate negotiations. In the UK, the new Labour Government, under Prime

Minister Tony Blair, adopted an ambitious voluntary target of a 20 per cent reduction in carbon dioxide emissions by 2010. In Germany, Environment Minister, Angela Merkel, who had chaired the negotiations in Berlin as the representative of the host country, renewed Germany's commitment to a 25 per cent reduction (Hatch, 2007). While other 'poorer and less green' member states such as Spain, Portugal and Greece, took a more cautious position, the UK and Germany were supported by other 'rich and green' states, such as the Netherlands, France, Denmark and Sweden (Ringius, 1999b: 17–18).[19] As a result, the UK and Germany continued to drive the EU position. Of course, as discussed in Chapter 3, their stance was made easier because greenhouse gas emissions in both countries were already decreasing as a result of economic restructuring in the early 1990s.

The leading role that the EU took in the climate change negotiations was engineered by the environment agencies in the member states, which continued to act as a progressive coalition as they had in Berlin – see Table 4.3. Consistent with the two-level perspective, this domestic coalition directly informed the behaviour of the EU. In the UK, for example, the new Deputy Prime Minister, John Prescott, was made Secretary of State for the Environment, Transport and the Regions, a 'super department', which under the auspices of the Deputy Prime Minister had significant clout in cabinet. Like his predecessor John Gummer,[20] Prescott was quick to argue in support of climate change action (Brown, 1997), and as UK officials pointed out, 'Prescott wanted to play a big role at Kyoto' and his seniority meant 'he was in constant contact with Blair' (EU-59). Prescott was also supported by key individuals who desired action on climate change, such as Peter Unwin, who headed the Global Atmosphere Division within the agency (EU-59; EU-44). In Germany too, the Federal Ministry for the Environment, Nature Conservation and Nuclear Safety (BMU) was one of the 'major protagonists in government efforts to formulate' climate change policy (Hatch, 2007: 45), where senior bureaucrats, such as Cornelia Quennet-Thielen and Franzjosef Schafhausen, were instrumental in developing the German position (EU-23; EU-32). As part of the troika, the Netherlands and its environment agency, the Ministry of Housing, Spatial Planning and the Environment (VROM), were also actively engaged (Kanie, 2003). In fact, during the 1990s VROM had more people working on the issue than any other ministry (Sewell, 2005: 101–2) and senior figures, such as Yvo de Boer and Bert Metz, played crucial roles in developing the Dutch and EU positions. Even in other environment agencies that did not have the same capacity there was a great deal of activity. For example, in France, the Environment Minister, (newly elected Dominique Voynet

Table 4.3 EU state actors 1996–1997

State actor	General position	Key figures
Council of Environment Ministers	Progressive	Johny Lahure (Luxembourg), Chair of the Environment Council
United Kingdom		
Department of Environment, Transport and the Regions (DETR)	Progressive	Secretary of State for the Environment John Gummer, 1993–1997 John Prescott, 1997–2001 Peter Unwin, Head of Global Atmosphere Division (DETR)
Department of Trade and Industry (includes responsibility for energy)	Cautious to regressive	Secretary of State for Trade and Industry Ian Lang, 1995–1997 Margaret Beckett, 1997–1998
HM Treasury	Cautious to regressive	Chancellor of the Exchequer Kenneth Clarke, 1993–1997 Gordon Brown, 1997–2007
Federal Republic of Germany		
Federal Ministry for the Environment, Nature Conservation and Nuclear Safety (BMU)	Progressive	Federal Minister for the Environment Angela Merkel, 1994–1998 Cornelia Quennet-Thielen, Head of Division, BMU
Federal Ministry of Economic Affairs (BMWi) includes responsibility for energy	Cautious	Günter Rexrodt, Federal Minister for Economics
European Commission		
DG Environment, Nuclear Safety and Civil Protection (DG XI)	Progressive	Ritt Bjerregaard, Environment Commissioner Director General, DG Environment Marius Enthoven, 1994–1997 James Currie, 1997–2001 Jorgen Henningsen, the Director responsible for climate change

Sources: Interviews and primary documents cited in text.

from the Green Party), made sure that the senior ranks of her agency were engaged (EU-40; EU-47).

As a group of 'true believers', as one former EU official labelled them (EU-49), many of these senior officials worked closely together to frame the EU negotiating position. Interviews indicate that the senior ranks of these agencies were bound by a belief in the science, which translated into support for action. As one British official put it:

> The underlying *fact* for the UK from day one was a clear belief in the science that it exists, and it is going to get worse. This was a cast iron belief (EU-56).

German, French, Dutch and Swedish officials made the similar statements about a 'scientific consensus' (EU-26). They also noted that environment officials often had good contact with many of their nations 'best' scientists who were members of the IPCC, which directly 'informed' their thinking (EU-32; EU-40; EU-52). This supports the proposition of the transnational perspective that the IPCC as an epistemic community helped to create a shared understanding based on its specialized knowledge.

It is no surprise then, that it was the environment agencies in the member states that adopted the most progressive positions in the inter-agency discussions. First, on greenhouse gas emissions commitments, environment agencies from the 'rich and green' member states typically argued for 'ambitious emissions reduction targets' (EU-59). According to British officials, 'the biggest tension', among agencies, 'was always how much we were willing to reduce greenhouse gas emissions' (EU-45). In the UK, the environment agency pushed for –15 per cent target on 1990 levels by 2010 for the EU, despite some opposition from other agencies (EU-59). BMU officials in Germany also advocated a –15 per cent EU target because they believed that 'European nations should lead on this as a union' (EU-32; EU-23). Fortified by the science, 'the true believers' in these agencies also accepted the principle of 'historic responsibility' and as a result, in contrast to the US, did not advocate new commitments for developing countries (EU-40).[21]

Second, on how emission reductions would be achieved, there was consensus among environment agencies on policies and measures, but not an emissions trading (EU-56). On policies and measures it would seem that most environment ministries, if not all, agreed with the French 'preference for policies and measures because it is well known, it is visible and the implementation is easy to understand' (EU-47). At the same time, it was acknowledged that policies and measures was a 'non-flyer' with the US, which, as I will discuss, would seem to have influenced the ultimate EU decision (EU-24). In France, there was much scepticism about the US proposal for emissions trading, as one official explained:

> It was because we had identified from the beginning that these types of systems would be full of loopholes and would enable many players to play with the system and this was raising significant implementation issues (EU-47).

However, not all environment agencies took this position. The Dutch 'were sympathetic to flexibility mechanisms as a means to keep costs down' (EU-52), and the British were 'very happy to support it as well' (EU-59). According to British officials, Environment Secretary, John Prescott, was more than comfortable arguing in support of emissions trading in EU discussions (EU-59). While these agencies still had reservations (Downie et al., 1998), in part, their support for emissions trading could be put down to their interactions with US officials outside of the domestic polity as the transnational approach would predict. As discussed in the last chapter, US officials actively sought to educate their European colleagues about the emissions trading system operating under the US Clean Air Act and this continued in the lead up to Kyoto.[22]

The power of this progressive coalition of environment agencies reflected the state of domestic political incentives and the power and distribution of opposing coalitions, as the two-level perspective would anticipate. First, the domestic political incentives worked in favour of the environment agencies. In many countries there was strong public support for climate change. In nations such as Germany and the Netherlands, parliamentary commissions, inquiries, and environmental NGO campaigns had been taking place since the mid-1980s (Kanie, 2003; Hatch, 2007). This helped to educate the public about climate change and the effect, as one German respondent describes, was to afford environment officials the political space to take ambitious positions.

> Because we had this consensus across German society, across different political parties it became easier to tell the public that we need things like targets and timetables (EU-23).

Second, whereas in the US a more regressive coalition was beginning to assert itself, this was not the case in the EU. Admittedly in Europe, those agencies that are the traditional counterweights to the environment agencies, such as the economic, energy and industry agencies did work in concert with each other. For example in the UK, Treasury and the Department of Trade and Industry (DTI), which also had responsibility for energy policy, 'were more or less aligned against the Department of Environment' (EU-59). However, at this point in the negotiations their engagement was limited and as a result they were not a powerful coalition in opposition to the environment agencies. For example, British, Dutch, French and German officials claimed 'there was relatively little engagement' from these agencies, especially at the senior levels (EU-36; EU-56; EU-52, EU-47; EU-32). As one former British Treasury official pointed out,

> The resources devoted to climate change in Treasury was one person at the most, it was generally a junior person, a desk officer (EU-36).

The only exception to this was the debate over the emissions target, which created the 'biggest tension' (EU-56). The concern from economic agencies, in particular, was that any international agreement be 'secured at minimal cost' (EU-36), because, as British, Dutch and French officials observed, there was a belief that environment agencies 'would do anything at any price' (EU-36; EU-52; EU-47). In France, this concern was more pronounced after Dominique Voynet became environment minister, as one official explained:

> … it was at the time when my minister had been appointed for two months and everybody was freaking out with the idea that the Greens would conduct the negotiations on climate change (EU-47).

Without a mandate to negotiate, the European Commission played a supporting role to the member states. The Environment Commissioner, Ritt Bjerregaard, was in charge of formulating the Commission's position, though because of the structure of the Commission, the policy work was done by DG Environment. At the time of Kyoto, James Currie was the Director-General and Jorgen Henningsen the director of the unit within DG Environment responsible for climate change (EU-49; EU-24). Some of these officials, most notably Henningsen, were part of the network of 'true believers', which had the support of the Environment Commissioner and formed part of the broader progressive coalition, which consistent with the two-level approach, informed the EU position (EU-52; EU-49).[23] Their aim was for the EU to adopt an ambitious emissions reduction to take to Kyoto. The target was –15 per cent by 2010, but it will become apparent, this met some opposition within the Commission (EU-42). The strategy of this network, who shared the same understanding of the science, 'was to keep climate change off the radar screen of other people in the Commission' in order to minimize likely opposition from other directorate-generals (EU-49).

For a time, this strategy proved successful, because, as they were in Berlin, the senior ranks of DG Energy, DG Industry and DG EcoFin 'were not very active' (EU-52). Most viewed climate change as little more than 'an exotic environmental issue' (EU-52) and 'there was very little knowledge of the issue' among the other DGs (EU-42). For example, one Commission official argued that 'DG Industry did not really know what was going on … and most people did not take the Kyoto process seriously' (EU-49; EU-61). DG EcoFin provided some technical assistance to DG Environment, but the senior personnel were indifferent. DG Energy was more engaged, but only in the lead up to Kyoto, before that it was seen as an 'esoteric exercise' (EU-51; EU-46).[24]

As a result of the limited engagement from other DGs, the 'true believers' in DG Environment were able to propose an EU target of –15 per cent by 2010. However, this could only be kept off the radar for so long. When DG Environment's proposal was finally put at the special chefs meeting, where members of each cabinet sit down to finalize proposals before they go to the Commissioners, the cabinets of other Commissioners were shocked, and the proposal met strong resistance. According to an official at the meeting, 'the most vocal opponent was the Industry cabinet', which almost always opposed proposals from the Environment Commission (EU-42). In the end, the Environment Commissioner, Ritt Bjerregaard, was able to get the College of Commissioners to propose a target of –15 per cent by 2010 to take to the Environment Council and the member states.

Yet because the Commission only played a supporting role, the critical discussions remained those in the member states. The clearest example was the debate on the greenhouse gas emissions commitment, which came to be known as the 'EU-bubble'. With the environment agencies in a few key member states and the Commission advocating an ambitious EU-wide target of –15 per cent by 2010, it was left to the Council of Environment Ministers to try to solve the size of the target and to decide how any emissions reductions would be allocated between member states (EU-52). After an earlier proposal by DG Environment was rejected by the member states at a workshop in Dublin in September 1996 during the Irish Presidency, the Dutch Presidency took the lead. In January 1997, the Netherlands, as the new EU President, released a report, which outlined the 'Triptique approach'. Commissioned by the Dutch environment agency, VROM, the report employed a sectoral approach to address the politically contentious issue of burden sharing where some member states would reduce their emissions and others would be able to stabilize or even increase their emissions. The report outlined four scenarios with EU-wide emission targets ranging from –9 per cent to –17 per cent on 1990 levels by 2010 (Ringius, 1999a).

However, some member states considered the targets too ambitious and a heated political debate commenced within the Environment Council. While the Danes, Dutch and the Germans, among others, continued to push for a –15 per cent target, others did not (EU-59; EU-25). As one European official recalled, 'too many member states were protesting that their share was too high, certainly Greece, Portugal, and Spain made a big fuss about this' (EU-52). To prevent a breakdown in negotiations,

Svend Auken, the Danish Environment Minister, engineered a compromise proposal whereby the EU target would remain at –15 per cent for Kyoto, but the burden sharing agreement would only cover a –10 per cent target. If the final agreement in Kyoto exceeded –10 per cent, the remaining emission reductions could be distributed at a later stage (Ringius, 1999a: 151).

Consistent with the two-level perspective that the distribution of domestic coalitions matters, the debate over the emissions target highlights the dominant role that a progressive coalition of environment agencies played in the lead up to Kyoto. Yet the interesting point, in the case of the EU, is that in the final instance it was the member states, largely informed by the environment agencies that made the call, not the Commission. A member of the EU Presidency at the time explained:

> There was no input really from the Commission on this analysis. It was at a time that the Commission was not given much room to manoeuvre, the Commission wasn't really trusted and the member states tried to keep burden sharing in its own realm. Commission proposals had not been resonating with member states and formally the Commission has no real power so the Commission had to earn their power by the value of their work. They were more like just one of the member states rather than playing any special role (EU-52).

Environmental NGOs and their Networks

While state actors in the member states and the Commission have been critical to determining the EU position at Kyoto, non-state actors also played a role. As in Berlin in 1995, the main environmental NGOs were the European Climate Action Network (EU-CAN), WWF, Greenpeace, and the IUCN among others, and they continued to argue for the EU to take a leadership role (EU-45; EU-26; EU-52).

Many of these NGOs continued to develop ties with government agencies, principally environment agencies, as part of a broad progressive coalition. In the UK and Germany, former environment officials confirmed that they met with these groups on a regular basis and had 'very good relationships with them' (EU-56; EU-32; EU-59). Also, the victory of centre-left governments in the UK and France in 1997 facilitated this process. This was especially the case in France, where the new Environment Minister, Dominique Voynet, from the Green Party, had a long association with the environment movement (EU-47). In addition, there were strong links between environment agencies and environmental NGOs in the Commission. A DG Environment official argued that 'NGO's were absolutely crucial', and in fact, DG Environment employed

Michel Racquet from Greenpeace to assist in preparation for Kyoto (EU-42; EU-49). Further successful campaigns by NGOs on other environmental issues, which had 'made commissioners sit up and take notice', led Ritt Bjerregaard, the Environment Commissioner, to pay greater attention to groups such as Greenpeace on climate change (EU-42).[25]

Environmental NGOs exploited these networks to share information with government officials and persuade them of the science of climate change. As several officials pointed out, NGOs 'have always been active and based on the science' (EU-52), and as a result they 'have played a very important role in helping the negotiators understand climate change' (EU-26; EU-28). Indeed, many of those interviewed had a high regard for the level of expertise within groups, such as EU-CAN and WWF, and acknowledged that they frequently came to rely on it (EU-56; EU-42). However, ties and information do not automatically equate to influence, and other studies have shown that while European environmental NGOs pushed the EU towards a progressive stance, many of their positions were not incorporated into the final agreement at Kyoto (Betsill, 2008; Corell and Betsill, 2008). One example is the opposition of groups such as Greenpeace and Friends of the Earth to emissions trading, which nevertheless was in the final agreement (EU-58; EU-30).

Yet there were some successful interventions, especially where environmental NGOs were able to infiltrate orthodox policy networks, as they had done in the US (Gough and Shackley, 2001; Rhodes, 2007). A case in point was the role of WWF and the development of the Triptique approach to burden sharing (Ringius, 1999a). While the Triptique approach was developed by the Dutch Presidency, it had its roots in an earlier report produced by WWF entitled *Policies and Measures to reduce CO_2 Emissions by Efficiency and Renewables* (WWF, 1996). WWF 'took the study to EU negotiators who liked it' and as a result Kornelius Blok was hired, along with other researchers involved in the report, by the Dutch Government to further the Triptique approach (EU-58; EU-52).

As the two-level perspective would predict, much like their American counterparts, European NGOs appear to have been most influential at the domestic level. Although groups such as EU-CAN, IUCN and WWF continued to send delegations to the international negotiations and coordinate their activity across borders, these activities were limited and there is little evidence to indicate that this influenced what the EU was prepared to sign, in contrast to what the transnational approach would anticipate (Betsill, 2008; Newell, 2000; Giorgetti, 1999). Instead, NGO officials often targeted their own national delegations at the international level, using mobile phones, for example, to call their national negotiators during closed door sessions.

Business Groups Play an 'Observer Role'

European business seems to have played a more muted role than environmental NGOs. Government officials claimed that business did not take the issue seriously at the time, and as one business representative put it, 'business was there but it was not involved' (EU-43; EU-39). There was 'a bit of change' in 1997, but for the most part, in contrast to groups in the US, business went along with the international negotiations and played more of 'an observer role' (US 8; EU 39). In other words, the international climate regime had still not triggered the type of domestic realignments among business groups that the international regime perspective would anticipate. Fossil fuel companies and associations such as BP, Shell and EUROPIA were most prominent, along with other industry associations like the CBI in the UK and the European-wide EURELECTRIC (EU-39; EU-45; EU-22; EU-24; EU-51). Other more progressive organizations included the World Business Council for Sustainable Development and the European Business Council for a Sustainable Energy Future (Lacasta et al., 2007).

Consistent with the two-level approach, the activities of these groups were largely confined to the national level where they believed they could have greater influence (EU-39). In doing so, business groups were more closely aligned with industry and energy agencies (Newell, 2000: Ch 5), though where these agencies were less engaged they also lobbied environment agencies (EU-24; EU-55; EU-39). A case in point was DG Energy in the Commission. A former energy official explained the relationship.

> Our deepest intellectual relationship was with the power sector and our biggest lobbying relationship was with the oil industry. … I was regarded as in the pocket of industry. But I was trying to bring a reluctant industry into the fold and for them to see climate change as legitimate. I interfaced with oil, coal and gas. There were regular consultations with business groups from these sectors. On the way to Kyoto the oil industry was beating me up, that was EUROPIA, but that was my community (EU-51).

The ties between energy agencies and business groups were often strengthened through working group consultations, which were regularly used by the British Government (EU-24; EU-39), and other joint initiatives. The same DG Energy official argued that one strategy employed to legitimate climate change among business groups was research collaboration (EU-51; EU-53). For example, between 1996 and 1997 a series of five workshops was organized by EURELECTRIC and DG Energy aimed at examining the role of electricity in fostering sustainable development. The workshops produced a joint 1997 report entitled *Electricity: The bridge between energy and sustainable development* (EU-51).

While US business groups, such as the GCC, were adamantly opposed to emission targets and timetables, European business was not. In May 1997, BP chief executive, Sir John Browne, delivered a speech at Stanford University in which he called for a precautionary approach in dealing with climate change (Browne, 1997). The speech marked an important shift in attitudes among oil companies, which had long been the loudest opponents to EU action. While officials in industry and energy agencies claimed that some parts of the business community remained 'hostile' to binding commitments (EU-36; EU-24; EU-47), there is little sense that business lobbying affected the EU position. In part, this likely reflects the limited engagement of these agencies themselves. On developing country commitments too, most business groups supported the application of binding commitments, yet this was not incorporated in the final EU position (Giorgetti, 1999).

While the evidence thus far supports the proposition of the two-level perspective that the COG monopolizes the external representation of the state and that the preferences of non-state actors are incorporated through the domestic polity, European business groups did operate across national borders and they often outnumbered environmental NGOs. For example, in one of the preparatory meetings to Kyoto in March 1997, there were approximately 35 industry groups with more than 150 representatives compared to 90 representatives from 25 environmental NGOs (Giorgetti, 1999: 220–21). Yet as interviewees pointed out, they 'rarely advanced the politics' in these negotiations (EU-37). In his study of business groups in the international climate negotiations, Newell (2000: 106) argues that the types of influence that business groups are able to exert at the national level are far less effective at the international level where delegates are more insulated from pressure groups. In short, there is scant evidence to suggest that European business influenced the type of agreement the EU was prepared to sign, especially via the type of networks that the transnational perspective would envisage.

Why did the EU Sign the Kyoto Protocol?

Since COP 1 in Berlin, the same coalition of environment agencies and environmental NGOs, supported by the Environment Council, which acted as the COG, dominated the discussions in the EU about the type of agreement the EU would sign in Kyoto in December 1997. This explanation is entirely consistent with what the two-level perspective would predict. However, as with the case of the US, the insights of the transnational perspective and the regime perspective do shed light on the role of some factors that are overlooked by the two-level approach.

Several factors are critical to understanding the dominance of this progressive domestic coalition. First, whereas in the US a more regressive coalition had emerged by 1997, this was not the case in the EU. Energy, industry, and economic agencies had yet to engage with the climate discussions at the senior level, and most business groups who traditionally sought to influence these state actors took on an 'observer role', monitoring the discussions, as opposed to the strident lobbying effort that took place in the US.

Second, the lack of participation from other agencies was in direct contrast to the active engagement of senior officials within the environment agencies. The network of 'true believers' were imbued with a shared belief in the science, which had been reinforced after the release of the IPCC Second Assessment Report in December 1995. As the transnational perspective would expect, the IPCC as an epistemic community worked to create a shared understanding about the issue among actors in this coalition. Together these actors sought to advance their agenda within their own governments and the Commission, while taking measures to minimize opposition from other agencies. Their position was also aided by the strong public support for action on climate change across political parties.

These two factors are essential to understanding the EU negotiating position. To be sure, the EU negotiating position at Kyoto was almost a direct reflection of the agreement nutted out between environment agencies. On greenhouse gas commitments, the EU target of a 15 per cent reduction by 2010 was what the environment agencies in the UK and Germany, among others, had advocated. It was also the position adopted by the College of Commissioners in the Commission, albeit after some push back from DG Energy and DG Industry to the original DG Environment proposal. Further, these agencies accepted the principle of historic responsibility and as a result did not argue for binding commitments for developing countries; and neither did the EU.

On the second set of issues, how emission reductions were to be achieved, there was a consensus in support for binding policies and measures, though not on emissions trading. While many environment agencies, notably the French, had serious reservations about the concept, others such as the British and the Dutch were more comfortable with market-based mechanisms. This reflected the differences among environmental NGOs as well. With very little engagement from economic and energy agencies on the issue, aside from the technical support of lower level bureaucrats, the ultimate EU position reflected a compromise between environmental interests. That is, that any emissions trading system should supplement domestic action.

However, to understand why the EU then went and signed an agreement at Kyoto that was outside its initial negotiating position, a third factor is crucial; the preferences of the Environment Council, or in the words of the two-level perspective, the preferences of the COG. As in Berlin, the Environment Council was determined to secure an agreement because of their domestic political imperatives. Many negotiators argued that the EU 'went to Kyoto wanting an agreement' (EU-37). The imperative was especially strong for national leaders of the newly elected centre-left governments in Europe.

> For Dominique Voynet. … If there was no agreement at Kyoto it would have been a political failure for her, being a green leader, coming back to France with failure on climate change would definitely have been an issue (EU-47).

To get an agreement, EU negotiators acknowledged that they had to 'get the US on board' (EU-23). As a result, the Environment Council was prepared to sign a protocol, which included an emissions trading system, albeit with the rules on the limits of its use to be deferred, which did not contain binding policies and measures, and one in which developing countries took on voluntary commitments, as the US had proposed with the so-called Article 10 (EU-59). As one negotiator put it 'we wanted a deal done in 1997 that is why we gave way' (EU-23; EU-20; EU-46). In the end, developing country opposition meant that the Kyoto Protocol did not include developing country commitments. Nevertheless, for the reasons outlined, the EU had been prepared to sign such an agreement. One that was different to the Berlin Mandate three years earlier and different to the agreement envisaged in their negotiating position.

CONCLUSION: HOW MUCH CAN THE THEORIES EXPLAIN?

The two-level perspective again best captures the behaviour of the US and the EU. In both cases progressive coalitions were able to dominate, despite opposition from some state and non-state actors. The preferences of the COGs, informed by the science and their own domestic political imperatives, intervened as necessary both at the domestic and international levels to secure the Kyoto Protocol. Nevertheless, there is evidence to support the propositions of the other theoretical perspectives. For example, as the transnational approach would expect, the IPCC as an epistemic community helped to create shared understanding among policymakers about the science of climate change, which underpinned the US and EU position. And, as it had in the lead up to Berlin, the international climate regime manipulated the cost-benefit calculations of the COGs by raising the cost of reputation. Importantly however, it is also evident that each of the perspectives does not sufficiently capture the temporal dimension of long negotiations and hence the factors behind the changes in US and EU behaviour, as I will discuss below.

Turning first to the negotiations in Kyoto, and consistent with the two-level perspective, the type of agreement the US and the EU were prepared to sign was largely determined by the power of a progressive coalition of government agencies and environmental NGOs. In the US, government officials in the EPA and the State Department, with the imprimatur of the White House as the COG, were largely able to determine the US negotiating position. In the EU, the 'true believers' in the environment agencies in the member states and the Commission were able to do the same, with, in many instances, the support of their national leaders. In both cases the scientific consensus that had been established with the release of the IPCC's Second Assessment Report and the resulting interaction between policymakers and scientists, encouraged by environmental NGOs, helped to fortify the position of these state actors within the national discussions.

However, as the two-level approach would predict, the power of this coalition reflected the distribution of other domestic coalitions. In the US, a more regressive coalition was beginning to emerge as were splits among state and non-state actors. This was evident with the divisions within the executive as the economic agencies became more engaged. It was also evident with the strident role played by Congress when in 1997, instigated by a network of business groups, the Senate passed the Byrd-Hagel Resolution, with the intention of limiting what the Clinton administration could ratify.

It was different in the EU. The progressive coalition of environmental interests did not face a regressive coalition of actors. In fact, EU negotiators did not face either of these two domestic constraints, at least not to the same extent. The natural counterweights to environment agencies in the EU, energy, industry and economic agencies were yet to engage at senior levels and most business groups, who traditionally were aligned with these state actors, were monitoring the internal discussions rather than lobbying against specific positions.

Yet the type of agreement the US and the EU were prepared to sign in Kyoto was different to that signed in Berlin three years earlier. For one thing, the US agreed to binding emissions targets and timetables, which it had opposed three years earlier. For another, the EU agreed to flexibility mechanisms, which it had also opposed in Berlin. To return to the central question of this book, that is, what factors lead a state to change its negotiating position and the type of agreement it is willing to sign, the limited explanatory power of each of the theoretical perspectives is evident.

Three factors can be identified that help to explain the changes in the position of the US and the EU. First, the level of engagement or mobilization of actors fluctuates over time. This was most evident in the US. In the early 1990s, those actors who would likely oppose the Berlin Mandate, the economic agencies and business groups, had not yet devoted sufficient attention and resources to formulate a position either for or against an agreement. By the mid-1990s this was changing, especially among business groups, which had actively mobilized to oppose an agreement at Kyoto. The result was a more restricted US position because of the pending ratification, which brought the US Senate into the debate as was clear in the position on developing country commitments. The increase in opposition is also directly related to the increase in calls for action. In other words, as actors push *for* an agreement it engages other actors to push *against* it.

Second, not only did the levels of engagement change, but so too did the networks between state and non-state actors. The two-level approach expects that the interests of non-state actors will be aggregated into a national foreign policy position, as it was in the US and the EU. However, in the course of prolonged negotiations the manner in which this occurs can change. The most obvious example is when a new state or non-state actor mobilizes. For example, the inauguration of President Clinton in 1993 triggered a distinct realignment between the White House and environmental NGOs. However, the networks can change even without a change in actor, and this is often overlooked by existing frameworks. For example, in the lead up to Kyoto, NGOs managed to manoeuvre in and out of orthodox policy networks to directly affect the type of agreement both parties were prepared to sign, as is evident with the examples of EDF in the US and WWF in the EU.

Third, the behaviour of the US and the EU seems to have been affected by changes in the state of expert knowledge among policy elites. Since the late 1980s a scientific consensus had begun to emerge that there was a discernible human influence on the climate. This was confirmed with the release of the IPCC Second Assessment Report in December 1995, which led to an increase in the interactions of scientists and policymakers. While it is difficult to measure the influence that this had on actors, interviews with policy elites suggest that these processes did affect their thinking and that of the COGs, as was evident in the public statements of government leaders, such as Al Gore and Tony Blair. In other words, as the scientific consensus on climate change permeated the policy elite, it encouraged them to develop negotiating positions and envisage agreements, which would reduce greenhouse gas emissions.

In summary, while the two-level perspective best captures the behaviour of the US and the EU at Kyoto, like the other perspectives, it fails to account for the dynamic nature of prolonged negotiations. In particular, it overlooks the fluid nature of state preferences in negotiations that drag on for years and sometimes decades. This will become more apparent in the next chapter on the negotiations in The Hague in 2000.

NOTES

1. The work of the Berlin Mandate was to be carried out by the so-called 'Ad Hoc Group to the Berlin Mandate' (AGBM), which was open to all parties to the UNFCCC (UNFCCC, 1995c). The AGBM held eight meetings between 1995 and 1997 in an attempt to resolve the slew of issues that needed to be reconciled if a legally binding agreement was to be possible in Kyoto (Earth Negotiations Bulletin, 1997).
2. JUSSCANZ refers to a bloc of developed countries that are not part of the EU. The members are: Japan, the US, Switzerland, Canada, Australia, Norway and New Zealand (Oberthur and Ott, 1999: 17–18).
3. Article 1 of the UNFCCC defines sinks as 'any process, activity or mechanism which removes a greenhouse gas, an aerosol or precursor greenhouse gas from the atmosphere' (UNFCCC, 1992). The term sinks is typically used to refer to greenhouse gas emissions removed by land-use change and forestry.
4. A 'non-paper' by Ambassador Estrada, which summarized the parties' positions, suggested growing convergence on some issues, such as emission budgets. See FCCC/CP/1997/CRP.2, 7 December 1997.
5. In the days that followed world leaders also heaped praise upon the Chairman. British Deputy Prime Minister, John Prescott, declared: 'This man has been brilliant. He's been able to create something out of nothing' (Boulton and Hutton, 1997). Interviews with US and EU officials, many years after the fact, confirmed the importance of his role. As one US negotiator fondly recalled: 'Everyone was passed out asleep on the tables, so there was not going to be any movement. Reporters had asked me what is going to happen and I said there is no deal between the US and the EU. But between 5am and 11am, Estrada had had a nap and he was ready for it. He stitched together a deal all by himself, it was unbelievable' (US-14).
6. I am grateful to Nathan Hultman for first drawing my attention to the importance of Article 10. Personal communication March 2010.

7. A textual analysis of news stories in the *New York Times* and *Washington Post* during the 1990s shows a spike in stories with the headline 'global warming' after COP 2 in Geneva (Agrawala and Andresen, 1999: 463).
8. For example, the Taskforce organized meetings between the President, Vice-President and Nobel winning scientists, briefed weather forecasters and invited them to televise from the Rose Garden (US-31), and in October 1997 it held a White House Conference on climate change attended by hundreds of politicians and civil society representatives (Oberthur and Ott, 1999: 69; Cass, 2006: 129–130).
9. The Byrd-Hagel Resolution resolved that: '(1) the United States should not be a signatory to any protocol to, or other agreement regarding, the United Nations Framework Convention on Climate Change of 1992, at negotiations in Kyoto in December 1997, or thereafter, which would: (A) mandate new commitments to limit or reduce greenhouse gas emissions for the Annex I Parties, unless the protocol or other agreement also mandates new specific scheduled commitments to limit or reduce greenhouse gas emissions for Developing Country Parties within the same compliance period, or (B) result in serious harm to the economy of the United States' (United States Senate, 1997b).
10. The idea, according to another US official, was that 'when the White House realized that the Byrd-Hagel Resolution was going to be passed it told all the Democrats to vote for it because it would send a message … that how could Senator Barbara Boxer and Senator Jesse Helms vote for the same thing and mean the same thing. … I don't know how true this is' (US-14). Other former White House and State Department officials claimed that the White House 'just kind of let it happen mainly because it was something that you could not vote against' (US-5). Another also pointed out that 'a lot of Democrat Senators voted for it, in deference to Senator Byrd' and because of his 'important role in the appropriations committee, so they wanted to stay on his good side' (US-4).
11. In the colloquies that accompanied the resolution, Senator Byrd gave qualified support to the science of climate change. He stated: 'In essence, the resolution accepts the thesis, which is still the subject of some dispute, that the increasing release of carbon dioxide and its accumulation in our atmosphere are causing a very gradual heating of the globe which has many adverse consequences for us all' (United States Senate, 1997a).
12. A former State Department official observed that prior to 1997 Congress had been quite dormant, 'but they woke up in the summer of 1997, with the Byrd-Hagel Resolution … this was a complete show-stopper' (US-14).
13. Eileen Claussen had been Director of Atmospheric Programs at the US Environmental Protection Agency, where she led the US delegation to the Montreal Protocol negotiations and had responsibility for the Clean Air Act.
14. Timothy Wirth left the State Department to head the United Nations Foundation (Cushman and Sanger, 1997; Helm, 1997). The reason for his departure varied among respondents. One claimed that: 'it was clear to everyone in the State Department that Timothy Wirth had been moved out by the White House because they did not want the point man for the US to fall under pressure as he had to the EU and Angela Merkel back in 1995 at Berlin' (US-37). Another argued that Strobe Talbott, then Deputy Secretary of State, decided that Wirth would not go to Kyoto because he did 'not like the optics' of having Wirth, who had already accepted a position at the UN Foundation, lead the US delegation (US-9).
15. Two months before the Kyoto negotiations, the *New York Times* reported that Robert Rubin and Larry Summers were meeting privately with the President to put forward their position on climate change, unbeknown to Vice-President Al Gore (Fineman and Breslau, 1997).

16. Cooper argues that the percentage reduction required by the Kyoto Protocol from projected emissions (business-as-usual) for the US is more than twice that which is required for the EU.
17. As one US official noted: 'You had a lot of political appointees coming in holding key leadership and management spots who wanted action on climate change' (US-27). Others pointed to interaction with US scientists, like those from NOAA, (US-9), the effect of which was to 'provide confidence that climate change was real' (US-7).
18. One senior negotiator explained the events: '[The target] was a product of the Vice-President. We talked about it briefly. … The White House said we can go –10 per cent… but Eizenstat who was connected with the NEC was not very interested in going very low or being very aggressive because he was aware of the concerns of the NEC. Through these discussions we sorted out that EU would have the most, US –7 per cent, Japan a little less because they were already quite efficient and then the rest of the targets fell into place' (US-9). This was confirmed in another interview (US-2).
19. The hand of the 'rich and green' states was strengthened by the fact that they comprised the troika (the Netherlands, Luxembourg and the United Kingdom) at the time of Kyoto, and hence led decision making of the Environment Council. In addition, as one EU official explained: 'The British were basically running the presidency when Luxembourg held it' (EU-24).
20. Several former British officials noted how closely Prescott had followed on from Gummer. For example, one stated about Kyoto: 'The UK never takes NGO people on delegations and so it was unprecedented to take John Gummer. But because of the seamless continuity between Prescott and Gummer he went along' (EU-56).
21. A former French environment official explained: 'My minister did not consider that binding commitments for developed and developing countries was reasonable. She said developed countries need to take commitments and take actions first and then with financial transfers they can trigger actions from developing countries' (EU-47).
22. For example, a French environment official stated: 'We had several coordination meetings before that including with the Americans. I remember I was sent to Washington to meet with people from the EPA and from economic think tanks a couple of months before Kyoto, September-October'. … And he continued, 'At the time we had lots of information coming from [French] economists that showed us that, yes, the US were indeed developing a large amount of studies in order to influence the position of other countries by showing that flexibility mechanisms were really the best thing' (EU-47).
23. A former EU official observed: 'The Commissioner for the environment at the time was converted on climate change, she may not have been actively engaged but she was encouraging this group' (EU-42).
24. As an example, a former DG Environment official described the difficulty he had getting DG Energy to participate in international meetings: 'I tried in vain to get them [DG Energy] to participate in the meetings so we would have one voice, but they just were not that interested (EU-25). This was confirmed by a DG Energy official: 'Initially we just saw climate change as a little disturbing, did we need additional problems, is it really happening? But from the run up to Kyoto this mentality began to change' (EU-21).
25. The example used by a former Environment Commission official was the Brent Spar incident (EU-42). In 1995, Greenpeace successfully campaigned against Shell UK over the disposal of the Brent Spar oil storage facility in the North Sea, generating a large public outcry against Shell.

5. From Kyoto to The Hague 1998–2000: shifting political dynamics and a question of ratification

COP 6 began in The Hague on 13 November, 2000. With 182 governments, 323 intergovernmental and non-governmental organisations, and 443 media outlets in attendance, some 7,000 individuals converged on the Dutch capital (Earth Negotiations Bulletin, 2000). Many hoped it would put the finishing touches on the Kyoto Protocol in preparation for ratification. However, in the months before the summit the sheer complexity of this task became apparent. Three sets of issues were under contention. The first was developing country commitments. Since developing countries had successfully opposed the inclusion of voluntary commitments (so-called Article 10) at Kyoto, the US had renewed its efforts to have them included. Two months before The Hague, Frank Loy, the lead US negotiator, urged parties to 'establish mechanisms that enable developing countries to voluntarily limit their emissions' (Loy, 2000). While the EU was not strictly opposed, developing countries remained vociferous in their opposition to any commitments (Earth Negotiations Bulletin, 2000).

The second set of issues was flexibility mechanisms. Article 6.1 of the Kyoto Protocol stated that the use of flexibility mechanisms 'shall be supplemental to domestic actions for the purposes of meeting [Kyoto] commitments' (UNFCCC, 1997a). In preparation for The Hague the EU called for a 50 per cent ceiling on the use of flexibility mechanisms in the first commitment period. The US was strongly opposed to any specific cap on the use of flexibility mechanisms and they were supported by other developed countries including Australia and Canada. The debate polarised into two camps: either the widespread use of flexibility mechanisms would compromise the Kyoto Protocol by opening large loopholes that developed countries could exploit to avoid taking domestic action to meet their targets; or the use of flexibility mechanisms would provide a cost-effective path for developed countries to realise their emissions targets.

The third set of contested issues related to carbon sinks, or what is technically referred to as land use, land use change and forestry (LULUCF). Article 3.3 of the Kyoto Protocol specified that activities from sinks could be used by developed countries to meet their commitments under the Protocol. The question for COP 6 was what activities could be counted towards a country's commitment. There were stark differences between the US and the EU. Whereas the US was pushing for a broad definition of sink activities to include all managed lands, the EU was adamant that there should be tight limits on sink activities and regarded anything less as a threat to the environmental integrity of the Protocol (UNFCCC, 2000).

The first week of negotiations made little headway in resolving these issues. By late in the second week the negotiations remained bogged down in the technical detail that had hindered progress in 1998 and 1999 (Earth Negotiations Bulletin, 2000). On Thursday 23 November, 24 hours before negotiations were due to conclude, COP President Jan Pronk distributed a special Note in an attempt to reconcile the key issues and provide the basis for a final agreement. The Note was divided into four 'crunch issues' and triggered frantic discussions as parties tried to reach agreement before the negotiations ended (Pronk, 2000). With time running out, developing countries were largely excluded from the closed-door meetings. In the final hours, only the US and the EU were 'left in the room', as the main parties – the US, UK, France and Germany – shuttled back and forward between meetings. However, after 36 hours stretching through Friday night and into Saturday the negotiations collapsed acrimoniously on 25 November (Earth Negotiations Bulletin, 2000 and interviews).

The popular interpretation of what happened was quickly splashed across the front pages of the international press; the Europeans had rejected a last minute deal struck between the US and the UK. American officials blamed the French and the Germans, while they, in turn, blamed the UK for risking the environmental integrity of the Kyoto Protocol by caving in to US demands (Fidler and Houlder, 2000). Personalities and recriminations aside, most explanations of the collapse at The Hague converge on two sets of issues: the negotiations over flexibility mechanisms and carbon sinks. The reason is simple; any changes to the rules governing these two areas had the potential to significantly affect a country's capacity to meet its Kyoto target. The political and economic ramifications of this, both domestically and internationally, were more than evident to the American and European negotiation teams.

Yet three years earlier, the US and the EU had reached an agreement. What had changed? What factors led the US and the EU to this decision? In this chapter I will again employ the two-level perspective, the transnational perspective and the international regime perspective to answer these questions. After looking at what type of agreement both parties were prepared to sign, I will turn first to the case of the US and then the EU.

WHAT TYPE OF AGREEMENT WAS THE US AND THE EU PREPARED TO SIGN IN THE HAGUE, 2000?

The breakdown at The Hague centred on the differences between the US and EU positions. While both parties had been able to agree to the Kyoto Protocol in 1997, they were not able to sign an agreement on the technical provisions of the Protocol three years later. As Table 5.1 shows, aside from similar positions on a compliance system, there were stark differences around developing country commitments, flexibility mechanisms and carbon sinks.

To begin with, in the course of the negotiations the US continued to insist on 'meaningful participation' from developing countries principally through voluntary greenhouse gas commitments (Stern, 1999). While the EU did not support this position, it signalled that it would not stand in the way (Earth Negotiations Bulletin, 2000). Rather it was on the remaining two issues – flexibility mechanisms and carbon sinks – that the US and the EU fell apart. The aim of the US was to reduce the economic costs of meeting its Kyoto emissions target. As Frank Loy, the lead US negotiator stated 'the final agreement must ensure that overall costs of compliance will be reasonable and no higher than necessary' (Loy, 2000). In contrast, the principal priority of the EU was to preserve the 'environmental integrity' of the Kyoto Protocol, as the chair of the Environment Council, Dominique Voynet, the French Environment Minister, had confirmed in the week before The Hague (European Report, 2000a).

The effect of these two conflicting priorities became apparent during the negotiations. Every attempt by the US to ensure that the rules of the Kyoto Protocol allowed the maximum use of flexibility mechanisms and carbon sinks was met with disdain by the EU. For example, the Swedish Environment Minister Kjell Larsson told the US that 'this is far beyond what is even possible to discuss' (Brown, 2000). Dominique Voynet went further arguing that 'we want to discuss ways in which we can meet our commitments, not ways in which we can avoid them' (Brown, 2000). US negotiators were equally critical. The State Department claimed that the

Table 5.1 US and EU positions at COP 6

Issue	US position	EU position	The Hague
Developing country commitments	Strongly advocates voluntary commitments for developing countries	Open to voluntary commitments for developing countries	No agreement
Flexibility mechanisms (emissions trading, CDM and joint implementation)	Strongly opposes any quantitative limit on the use of flexibility mechanisms	Strongly advocates a 50 per cent ceiling on the use of flexibility mechanisms to meet emissions targets	No agreement
Carbon sinks (otherwise known as land use, land-use change and forestry)	Proposes broadening the definition of sink activities to include all managed lands	Opposes broadening the definition of sink activities	No agreement
Compliance system	Advocates legally binding consequences for exceeding emissions targets	Advocates legally binding consequences for exceeding emissions targets	No agreement

Sources: US Department of State, 2000; Earth Negotiations Bulletin, 2000; European Environment Council, 2000; Loy, 2000; and interviews.

EU's push for a quantitative limit on emissions trading was 'an attempt to rewrite the Kyoto Protocol' and that the EU wanted to 'restrict every flexibility mechanism agreed to at Kyoto other than its own [the EU bubble]' (Foley, 1999). In the end, neither party was willing to budge, and agreement could not be reached.

THE CASE OF THE US

State Actors: 'How are we going to get this Ratified?'

The period 1998 to 2000 represented the final years of the Clinton administration. Since the beginning of their first term in office in 1993, the President, and especially the Vice-President, had actively pushed the climate change agenda. However, by 1998 the White House was increasingly constrained by a hostile Congress as debate focussed on the economic implications of implementing the Kyoto Protocol. It, together with the EPA, the DoE, the economic agencies and the State Department, remained the principal actors in the inter-agency discussions – see Table 5.2.

For the White House the main issue was, as one former official put it, 'how are we going to get this ratified?' (US-34). While the White House had long argued for action on climate change, by the late 1990s the politics were unenviable. This was manifest in two ways both of which support the propositions of the two-level approach. First, the preferences of the COG (the White House) were increasingly determined by a desire to affect its domestic political position. This was especially so for the Vice-President, who in the lead up to The Hague, was becoming 'concerned about his electoral prospects' (US-19). Former officials claimed that Al Gore's presidential aspirations constrained the ambition of the White House and many of his advisors wanted to downplay the issue of climate change because they believed it would be 'a loser issue in key industrial swing states' (US-1; US-19; US-5; US-4).

Second, and also consistent with the two-level proposition that the COG largely responds to domestic political pressures, such as domestic coalitions, the White House was forced to respond to an emerging veto coalition. The first arm of this regressive coalition was the Republican controlled Congress.[1] Since the agreement in Kyoto in December 1997, the Congress had actively sought to frustrate White House efforts to ratify or implement the provisions of the Kyoto Protocol. For example, Republicans attached an amendment to the Fiscal Year 1999 Appropriations Bill stipulating that no funds could be used for Kyoto related initiatives. The Congress also subpoenaed government agencies and held up confirmations to thwart the work of administration officials, as was the case with Frank Loy, Under Secretary of State for Global Affairs (Royden, 2002: 450–51).

According to officials in the EPA and the DoE many of these attempts were successful.

> Everyone in government became concerned that anything they put in their budget with the words climate change would be killed by the congressional appropriation committee, so with every budget document we were doing ctrl find [on the keyboard] to make sure climate change was not in there. … So instead of saying climate change in a document we would say we want renewable energy projects not because of climate change but because of energy security (US-11).

Table 5.2 US state actors 1998–2000

State actor	General position	Key figures
White House	Progressive	President, Bill Clinton Vice-President, Al Gore
		Richard Ballantine, Deputy Assistant to the President
Council of Environmental Quality (CEQ)	Progressive	Chair of CEQ Kathleen McGinty, 1993–1999 George Frampton, 1999–2001
National Security Council (NSC)	Progressive	Ian Bowles, Director for Global Environmental Affairs at the NSC
White House Climate Change Taskforce	Progressive	David Gardiner, Executive Director
EPA	Progressive	Carol Browner, EPA Administrator
		John Beale, Deputy Assistant Administrator
		David Doniger, Director, Air Office
DoE	Cautious	William Richardson, Energy Secretary
		Richard Bradley, Director, Office of Global Environment
		Abraham Haspel, Assistant Deputy Under Secretary
'Economic agencies'	Cautious to regressive	Secretary of Treasury Robert Rubin, 1995–1999 Lawrence Summers, 1999–2001
		Deputy Secretary of Treasury Lawrence Summers, 1995–1999
		Gene Sperling, Director of the NEC Chair of the CEA Janet Yellen, 1997–1999 Martin Baily, 1999–2001
State Department	Progressive	Under Secretary for Global Affairs Ambassador Stuart Eizenstat, 1997–1998 Frank Loy, 1998–2001
		Ambassador Mark Hambley, Special Negotiator for Climate Change
		David Sandalow, Assistant Secretary of State for Oceans and International Environmental and Scientific Affairs

Sources: Interviews and primary documents cited in text.

Further, because the Kyoto Protocol could not be ratified without the consent of the Senate, its twin concerns as expressed in the Byrd-Hagel Resolution – the role of developing countries and the potential economic costs – became the dominant issues for the White House. First, as Todd Stern stated on multiple occasions, the US was 'not … intending to submit the treaty for Senate ratification until' there was a 'meaningful level of participation from developing countries' (Stern, 1999). The White House tasked the State Department with encouraging developing country participation. In 1998, Secretary of State, Madeleine Albright, announced a diplomatic 'full court press' to encourage meaningful developing country participation (Albright, 1998). Frank Loy, who replaced Stuart Eizenstat in 1998, was to lead the agency's efforts. With a series of 'charm offensives' the State Department sought to persuade major developing countries to voluntarily adopt greenhouse gas emissions targets, which they had failed to do at Kyoto (Doran, 2000). The diplomatic strategy, which in the words of the two-level approach was an attempt to manipulate the international level to appease a domestic coalition, showed signs of success at COP 4 in 1998 when Argentina, and then Kazakhstan, split from the G77 and announced that they would adopt voluntary targets (Pitzl and Stewart, 2000). The US response was immediate. The Vice-President said he was 'particularly pleased with the growing engagement of developing countries' and the US was spurred to sign the treaty (Royden, 2002: 457).[2] However, the Senate remained unconvinced, and as other developing countries failed to follow suit, the White House and State Department were left to ponder whether developing countries would again be left out of any agreement.

Second, the 'administration quickly realised that an informed public debate over the Kyoto Protocol could not proceed without an economic analysis of the agreement' (Aldy, 2004: 95). The economic agencies were given the task of assessing the economic costs. Former administration officials argued that the Treasury, the CEA and the NEC 'were all highly involved' in the inter-agency discussions in 1998 (US-23; US-19; US-34). With growing pressure from Congress, 'the view from the White House' was that the administration 'had to testify before Congress about what the economic impacts would be' (US-23). The task was given to the CEA (US-23; US-26). Janet Yellen chaired the CEA and she became the administration's spokesperson on the economics (Aldy, 2004). The challenge for the CEA was to develop an economic model that would show that the US could meet its Kyoto target at a manageable cost. The problem was that many in Treasury including Deputy Secretary Lawrence Summers, who replaced Robert Rubin in 1999 as Secretary, did not think this was possible.

> Larry Summers thought abatement and the Kyoto Protocol were very hard to do and required heavy economic lifting. Summers convinced Bob Rubin that

> reducing greenhouse gas emissions was so pervasive that it was going to be very expensive (US-4).

While the CEA 'shared these doubts' (US-4), they nevertheless began to model the cost of emission reductions under different scenarios. Early attempts suggested that the cost would be high, yet as will be discussed, other agencies, notably the DoE, were arguing otherwise. This was not the view within the economic agencies (Frankel, 2001). As the pressure mounted for the CEA to testify before Congress, discussions among economic officials became heated as the positions of individuals oscillated. Even Janet Yellen was reluctant to testify that the US could reduce emissions by 7 per cent on 1990 levels at a manageable cost.[3] The CEA report was ultimately finalised after an 'endless back and forth with the White House about the wording' (US-23).[4]

In July 1998, the CEA tabled the report, the findings of which were also presented and scrutinised in Yellen's ten testimonies before Congress between 1998 and 1999. The report concluded that 'with the use of flexibility mechanisms', the US 'can reach its Kyoto target at a relatively modest cost' (Council of Economic Advisors, 1998: i). The modelling suggested that the cost would be $7–$12 billion per year in 2008 to 2012, or 0.1 per cent of projected GDP. It predicted that emission permits in 2010 would cost between $14 and $23 per ton of carbon equivalent (Council of Economic Advisors, 1998).

Much attention focussed on the underlying assumptions of the model, especially from critics who argued that the results were unduly optimistic (Toman, 2004). Two key assumptions drove the analysis: full international emissions trading and 'meaningful participation' from developing countries. Although reported only implicitly rather than explicitly for political reasons, it was estimated that without full international trading that included developing countries, the costs of meeting Kyoto would be 5 to 8 times higher and hence the prices of emission permits per ton of carbon equivalent would rise to in excess of $100 (Frankel, 2001). The political likelihood of these two conditions being met was not discussed in the report, although as Yellen had testified before Congress in March:

> … while our analysis may be predicated on some ambitious conditions concerning trading and developing country participation, it is exactly those conditions that form the foundation of the US position in international negotiations (Yellen, 1998).

Interestingly, the CEA analysis 'made no allowance' for the role of sinks in removing carbon from the atmosphere, though as a member of the CEA pointed out, this was because of the technical difficulties in doing so (Frankel, 2001: 12–13). However, the economic agencies, like almost all agencies, advocated the broad use of sinks to reduce the economic impact of emission reductions (US-7). While the US position on sinks was to become a critical fault line with the EU, there is little evidence to indicate that it was a point of contention within the inter-agency discussions.

The CEA estimates became the reference point for inter-agency disputes over the costs of the Kyoto Protocol (Frankel, 2001; Royden, 2002). It was not just the economic agencies that were looking at the economic costs. The DoE was too, both within the agency proper, and in the Energy Information Administration (EIA), an independent statistical agency under the umbrella of the DoE. As noted in the last chapter, the DoE had earlier released the 'five-labs' study, which concluded that emissions could be reduced at a very low cost (Aldy, 2004). CEA officials had dismissed the findings as the claims of 'techno-optimists' (Frankel, 2001).

One year later in October 1998, the EIA released a report suggesting that 'significant increases in energy prices may be required' for the US to meet its emissions reduction target (Energy Information Administration, 1998a). The findings of the report were at odds with those of the CEA. It estimated that in 2010 emission permits would cost around $348 per ton of carbon if the US was to meet its Kyoto target (Energy Information Administration, 1998b). In comparison, the CEA had estimated that emission permits in 2010 would cost between $14 and $23 per ton of carbon equivalent (Council of Economic Advisors, 1998). As the *New York Times* reported: 'the new study plays directly into the hands of treaty opponents' (Cushman, 1998). And so it did. For the White House too, the EIA analysis was 'bad news' (US-4).

The report also suggests many Republicans in Congress viewed the DoE as an ally within the administration. In fact, the EIA report had been commissioned by the House Science Committee, whose chairman, Republican James Sensenbrenner was a leading critic of the Protocol. With the House Committee specifying that the EIA modelling make assumptions very different to those of the CEA analysis, including an absence of full international trading and the participation of developing countries, it was sure to find that the economic impact on the US would be ruinous (US-23).

Further, former DoE officials revealed that while officially they were pushing for agreement at The Hague, because 'the White House wanted an agreement', their predisposition was against an agreement (US-2). At a dinner at the house of the American Ambassador in The Hague in November 2000, a DoE official recalled how Republican Senator Chuck

Hagel had said to him and others, 'I am glad there are DoE people here so that we won't sign an agreement that will harm our economy' (US-2). Senator Hagel, who was on the delegation at the time, had been a vocal critic of the Kyoto Protocol and, as discussed in the last chapter, was the reputed 'point man' for the fossil fuel lobby, the GCC; and as such was part of the regressive domestic coalition.

The EPA was also active in the inter-agency discussions. With Carol Browner still at the helm, the EPA remained a staunch advocate of action on climate change. According to former officials, the EPA supported State Department efforts to encourage developing country participation and, with the ongoing success of the sulphur dioxide regime (see Chapter 4), it supported the US position on emissions trading to reduce the economic cost of emission reductions (US-22). In addition, EPA officials continued to participate in transgovernmental interactions to coordinate policy with their EU counterparts about emissions trading, as the transnational approach might expect (US-6).

However, by 1998 the EPA had lost some of its influence within government. As the discussions focussed increasingly on the economic impacts of the Kyoto Protocol, the economic agencies became a formidable counterweight to the positions put forward by the EPA and officials noted the 'tensions' between the agencies (US-21). This was compounded by the restrictions Congress had imposed on the EPA's agenda (US-22), as part of its broader attacks on the Clinton administration. For example, in 1999, House Resolution 4194 – the appropriation bill for the EPA – contained a ban on spending for any measure to implement the Kyoto Protocol including measures designed to educate the public about the issue of climate change (Bryner, 2000: 118).

Throughout the period 1998 to 2000, Congress also maintained its attacks on the science of climate change (Bryner, 2000). Nevertheless interviews with administration officials, including with the members of the economic agencies, indicate a firm belief in the scientific consensus (US-23; US-6; US-10). This supports the proposition of the transnational perspective that epistemic communities promote organizational learning by helping to create shared understandings about a problem. Indeed, their awareness of the problem seems to have become more acute as the scientific reports mounted. For example, a former White House official noted the ongoing interaction with NOAA scientists, such as James Baker, who often briefed White House staff and the Vice President, 'who was very much a sponge on the science' (US-5). While the inter-agency discussions concentrated on the domestic political debates around economic costs, the policy elites who participated in them remained bounded by the science.

Environmental NGOs: 'Stick with the Kyoto Protocol'

Non-state actors, such as environmental NGOs and business groups, played important roles in the lead up to Berlin and Kyoto. It was no different in the lead up to The Hague.

In the late 1990s, the leading environmental NGOs in the US maintained close links with the Clinton administration. The EDF, NRDC, and the Sierra Club, among others, regularly met with administration officials. For example, interviewees recalled that George Frampton, who replaced Katie McGinty as head of CEQ, and David Sandalow at the State Department had meetings with the CEOs of these environmental NGOs prior to The Hague (US-34; US-37). Further, many people who had moved into senior positions in the administration had spent time in these organisations. For example, David Gardiner, who had moved from the EPA to become Executive Director of the White House Climate Change Task Force, had spent 15 years at the Sierra Club (US-7; US-19). The same was true of others in the White House, the State Department and the EPA. In short, these networks were a central part of a progressive coalition of environmental interests in the lead up to The Hague.

After the agreement in Kyoto, and the rise of domestic opposition in Congress, NGOs focussed on ensuring that the administration would 'stick with the Kyoto Protocol' (US-34). NGOs also participated in a campaign to ratify the Protocol by 2002, which gained recognition in the international negotiations (Royden, 2002: 465). For example, John Passacantando, Executive Director of Ozone Action, argued that it is 'anti-American and counter-productive' for the US to wait for developing country commitments before moving ahead with Senate ratification. Spokespeople for the NRDC and the American Council for an Energy-Efficient Economy made similar calls, and targeted business groups, such as Chrysler, which advocated delay (Pitzl and Stewart, 2000: 7). As I will discuss in the next section, the efforts of NGOs to dilute the influence of business groups that advocated abandoning the Kyoto Protocol was particularly successful.

However, it is often difficult to ascertain how much influence NGOs have on the position of states. On the surface, NGOs were successful in ensuring that the administration did not walk away from the Kyoto Protocol. Part of the strategy of NGOs was to target the Vice-President; NGOs bought full page advertisements in national newspapers 'saying Al Gore was not green enough' (US-34). According to former administration officials it seemed to work:

> Al Gore wanted the environment groups on side during the primaries and during the election and he could not let a big gap open up. So because they were saying 'stick with the Kyoto protocol' so was he (US-34).

Yet the Vice-President's political aspirations cut both ways. As the national election campaign gathered pace in 2000, Al Gore was also careful not to be seen as 'too green'. A former White House official explained:

> The Vice-President was thinking that they had delivered to the environmental constituency with Kyoto, but if he pushed too hard he would alienate some of those key swing states (US-1).

Consistent with the two-level approach, this evidence highlights how non-state actors seek to influence the preferences of the COG, given the ultimate effect such preferences have on state behaviour. While in this case the NGOs were successful in forcing the administration to stand by the Protocol, as the public debates shifted to the economic costs associated with implementation, the influence of NGOs seems to have waned. The actions of environmental NGOs in this period provide little support for the propositions of the transnational perspective or the international regime perspective. For example, there is little evidence to suggest that the transnational networks that did exist affected US behaviour, or that that the international regime was influencing the behaviour of environmental NGOs, at least so far as can be identified.

Business Groups on the Hill Preventing Ratification

After Kyoto, the regressive elements of the business community that had previously opposed a Protocol now turned their attention to preventing it from being ratified. The main players included Exxon and Mobil, Edison Electric, API and the GCC, among others (US-19). However, by the late 1990s the business community was also splintering as more companies began to accept the science and recoil from outright rejection of climate change action. As will become clear, the fractures among business groups were encouraged by environmental NGOs, although it is likely that the international climate regime also played some role in triggering such realignments, as the regime approach would predict.

To make their presence felt business groups continued to network with government agencies. For the most part this meant the DoE, though they also met with White House and State Department officials (US-5; US-37). One Treasury official quipped, the 'DoE is like the commerce department for the energy industry' (US-13). Yet for the most vocal opponents of the Protocol, such as the GCC, the general feeling was that the Congress was a better option than the Clinton administration:

> We did a little with DoE, but most of our work was on the Hill, we did not deal a lot with government. DoE was not going to help us, nor was this administration (US-28).

The strategy of these groups was to reframe the public debate in order to cast the Kyoto Protocol as unfair, given of the absence of commitments for developing countries and the potential economic costs. In tandem with Republicans in Congress, groups such as the GCC and the API devoted money and resources to trashing the Kyoto Protocol and emphasising these two points (US-19; US-37). The strategy worked, according to administration officials, as the 'opposition managed to define "Kyoto" as a bad word, simply with money' (US-6). A former White House official reflected on the public debate in 1998:

> I don't think there was ever an attempt to make the case that it is fair for developing countries not to have a target. … In a way, we may have allowed the opposition to get out ahead of us on this issue of fairness (US-1).

Other researchers have shown that the actors that succeed in establishing a frame that is consistent with their preferences are more likely to achieve their objectives (Sell, 2003; Joachim, 2003). This seems to have been the case in the US. Business groups continued their attempt to frame the public debate, and to the extent that they were able to prevent the Clinton administration from ever sending the Kyoto Protocol to the Senate for ratification, the climate negotiations in the US seem to be further evidence of the influence of framing.

However, business groups opposed to the Protocol did not have it all their own way. By 1998 environmental NGOs had mounted a rearguard action to, in the words of the two-level approach, shift the power and distribution of the the regressive coalition. The Pew Center on Climate Change was a crucial part of this. Established in 1998, it sought to pressure the business community, including some of the world's most powerful corporations, who had been members of the GCC to defect and join the Business Environmental Leadership Council, which took a more constructive approach to the climate change negotiations (Gough and Shackley, 2001; Carpenter, 2001; Levy, 2005). It was quickly successful. After the withdrawal of BP from the GCC in 1997, others followed including Shell, Ford, Daimler-Chrysler, and Texaco (US-1; US-5). Business lobbyists interviewed blamed the Pew Center:

> The Pew Center miraculously popped up when the GCC was being very successful. They targeted individual companies and there were shareholder resolutions saying they wanted this company to leave the GCC and companies were worried about their reputation so they started dropping out (US-28).

Crucially, the Pew Center initiative 'opened up doors for progressive democrats who previously thought that there was no business support on climate change' (US-5). However, it could equally be argued that this initiative was too late to change the US position before The Hague negotiations commenced. In response to questions about the decline of the GCC, a number of business representatives happily pointed out that:

> The GCC achieved its objectives, there is no Kyoto protocol, no domestic legislation. The GCC was only designed to kill Kyoto and that was the game (US-28).

To the extent that it is possible to determine the influence that business groups had on the type of agreement that the US was prepared to sign, it is evident that their most important actions, like those of NGOs, occurred within the domestic polity, as the two-level approach would predict. While business groups were not the only actors seeking to prevent the White House from ratifying the Protocol, they were an important part of an emerging veto coalition. By highlighting the role of developing countries and potential economic costs, it would seem that business groups helped to shape the administration's approach in The Hague. Many of these groups also sent large delegations to the international negotiations in The Hague, but as business representatives were quick to point out, when there they mainly targeted the US delegation (US-33; US-5). In other words, there was no evidence of transnational networks, at least no networks that influenced what the US was willing to sign.

Why did the US Not Sign an Agreement in The Hague?

In the lead up to The Hague, a regressive coalition of state and non-state actors had come to dominate discussions in the US. As attention turned to the question of ratification and the likely economic impacts of the Kyoto Protocol, divisions within the US Government, which had emerged in the lead up to Kyoto in 1997, intensified. Splits within the executive branch were more apparent as the economic agencies engaged in the inter-agency discussions, and further, the earlier tensions between the executive and the Congress had become entrenched by 1998. These shifting domestic dynamics combined with the changing preferences of the White House, notably the Vice-President's concern for his electoral prospects, ensured that the US was less willing to search for an agreement.

The two-level perspective best captures the significance of this distribution of domestic coalitions and its impact on US behaviour. The growing power and influence of a regressive veto coalition, in particular, was evident in the US negotiating position. As discussed, the two key planks of the US negotiating position were developing country commitments and flexibility mechanisms. The US position advocated voluntary commitments for developing countries and the use of flexibility mechanisms, such as emissions trading, without any quantitative cap. This position was a direct reflection of that advocated by the economic agencies. In fact, these were the two key assumptions that allowed the economic modelling by the CEA to conclude that the US could meet its commitments under the Kyoto Protocol at a 'relatively modest' cost. Whereas in 1996–97 the senior ranks of the economic agencies had only started to engage in the debate, by 1998 they were active players and, as a result, their interests were more fully incorporated into the US negotiating position.

However, the increasing level of engagement from economic agencies does not completely explain the US negotiating position, or the decision not to sign an agreement in The Hague, as it had three years earlier at Kyoto. Three factors would seem to be important that support the principal propositions of the two-level approach. First, by 1998 a regressive coalition in the Republican controlled Congress in tandem with powerful business groups had successfully framed the domestic political debate in their terms. As a result, any agreement that did not include commitments for developing countries and sufficient scope for the US to use flexibility mechanisms to meet its Kyoto target at a reasonable cost, would have been politically explosive for the Clinton administration.

Second, the preferences and strategies of the COG had changed since Kyoto. In the context of the democratic primary campaign and the 2000 national election, Vice-President Al Gore was much less inclined to push the issue of climate change as he had been in the early 1990s. As former administration officials pointed out, the Vice President and his advisors were concerned that if the White House took an ambitious position in the lead up to The Hague it would likely be 'a loser in key industrial swing states'. Furthermore, the White House, and the Vice-President in particular, believed that with the agreement in Kyoto they had already delivered to their environmental constituency, namely environmental NGOs. As long as the administration did not walk away from the Protocol, the feeling within the administration was that they would be able to hold onto the environmental community.

Third, at this stage of the negotiations the influence of the international climate regime no longer appears to have manipulated the cost-benefit calculations of the COG by raising the cost of reputation. The Clinton administration was not concerned about being seen as a pariah on the international stage. Instead, by the end of the second term, the greater concern for Gore was his domestic reputation in the context of a national election campaign. Further, according to a senior official, 'the thinking of the administration was that if the Senate is going to reject Kyoto that is their business' but having signed the Kyoto Protocol they were already 'on the right side of history' (US-34). None of these factors meant that the US did not want an agreement, it simply meant that by 2000, 'no deal' was, as David Sandalow of the State Department put it, 'better than a bad deal' (Herbert, 2000).

THE CASE OF THE EU

State Actors: Green Politicians and Economic Bureaucrats

With a mandate to negotiate, the Environment Council continued to lead the EU in the climate negotiations. Environment agencies in the member states, therefore, remained the key actors in the inter-agency discussions along with the economic, energy and industry agencies. This was evident in the EU discussions on the burden sharing agreement, which had to be re-negotiated after the Kyoto Protocol. Member states took the lead with little input from the Commission.[5] However, the Commission did play a role on other issues and principal actors in the Commission paralleled those in the member states with DG Environment, DG Energy, DG Industry and DG Economic and Financial Affairs all involved – see Table 5.3.

Following Kyoto, the Environment Council experienced a significant change in its political dynamics. A series of national elections in the 15 member states had catapulted the Green Party into coalition governments. The result was a large shift in the composition of the Environment Council. By the time of The Hague in November 2000, five of the 15 environment ministers (from France, Germany, Italy, Finland and Belgium) were Green Party members including the chair, French Environment Minister Dominique Voynet. The Green ministers together with supportive environment ministers in the Netherlands and Denmark were able to dominate the Environment Council discussions. A former environment minister explains:

Table 5.3 EU state actors 1998–2000

State actor	General position	Key figures
Council of Environment Ministers	Progressive	Dominique Voynet (France), Chair of the Environment Council
United Kingdom		
Department of the Environment, Transport and the Regions (DETR)	Progressive	John Prescott, Secretary of State for the Environment Peter Betts, Head of Global Atmosphere Division (DETR)
Department of Trade and Industry (includes responsibility for energy)	Cautious to regressive	Secretary of State for Trade and Industry Margaret Beckett, 1997–1998 Peter Mandelson, 1998–1998 Stephen Byers, 1998–2001
HM Treasury	Cautious to regressive	Gordon Brown, Chancellor of the Exchequer
Federal Republic of Germany		
Federal Ministry for the Environment, Nature Conservation and Nuclear Safety (BMU)	Progressive	Federal Minister for the Environment Angela Merkel, 1994–1998 Jurgen Trittin, 1998–2005 Karsten Sach, Head of Division, BMU
Federal Ministry of Economics and Technology (BMWi) (includes responsibility for energy)	Cautious	Werner Müller, Federal Minister for Economics and Technology
European Commission		
DG Environment, Nuclear Safety and Civil Protection (DG XI)	Progressive	Environment Commissioner Ritt Bjerregaard, 1995–1999 Margot Wallström 1999–2004 James Currie, Director General, DG Environment Jos Delbeke, Head of Unit, climate change

Sources: Interviews and primary documents cited in text.

> The five Greens coordinated their position prior to Council meetings … on top of the five there was also Jan Pronk from the Netherlands and Svend Auken from Denmark. So you almost have a majority, because you have seven of 15, but also included in the original five are France and Germany, who have more votes, because votes are based on per capita. So we were in a very good position (EU-19).[6]

This new political reality ensured that the Environment Council in coalition with environment agencies would pursue a strong climate agenda. A week before The Hague, the Council convened a special climate change session at which it confirmed that the principal priority of the EU was to preserve the 'environmental integrity' of the Kyoto Protocol (European Report, 2000a). Environmental integrity was to be the dominant theme of the Environment Council in the aftermath of Kyoto. For many environment ministers and agency officials a focus on environmental integrity stemmed directly from a strong understanding of the climate science. German, Belgium, and Swedish officials all claimed that they tried to push for positions 'that were as close as possible to what the science was calling for' (EU-32; EU-19; EU-26). Many pointed to the 'close contact' that they had with their national scientific institutions and scientists (EU-32). The growing consensus on the science was also employed by ministers and senior bureaucrats to defend their positions in inter-agency disputes (EU-32; EU-35). This evidence supports both the two-level perspective, to the extent that the preferences of the COG (the Environment Council) helped to shape the EU position; and it supports the proposition of the transnational perspective, to the extent that scientists as part of an epistemic community helped to establish a shared understanding around the science.

In this political context, the environment agencies maintained a prominent role in the inter-agency discussions. In Germany, the BMU, which was now led by Jurgen Trittin of the Green Party, played a lead role determining the German position. The BMU was also often supported by the Chancellery in inter-agency disputes. According to a BMU official, Trittin's success in resolving the nuclear issue in Germany 'increased his standing within the Government' and put him in regular touch with Chancellor Schroeder (EU-23). In the UK too, Deputy Prime Minister John Prescott remained Secretary of State for the Environment, Transport and the Regions, and this 'super department' continued to be a powerful player in inter-agency debates (EU-44). As discussed in the last chapter, in most of the key member states, including France and the Netherlands, the senior ranks of environment agencies were active.

However, there was one significant change after 1997 within the progressive coalition, which would alter its preferences. In some agencies, the 'true believers' were replaced by economically-minded bureaucrats, who took a different approach to developing climate policy. This was especially evident in DG Environment. After Kyoto 'a wholesale change' took place within the agency (EU-24). Jorgen Henningsen, the director responsible for climate change, left the agency along with many of his staff who had helped formulate the Commission's position prior to Kyoto (EU-24; EU-49). The space was filled by Jos Delbeke, who had a PhD in economics and had worked for some years in the economic instruments unit of DG Environment. He was joined by other economists with training in market-based instruments including Ger Klassen, Peter Vis and Peter Zapfel (Skjaerseth and Wettestad, 2008: 74). Although these changes were not orchestrated against the 'true believers', they were to have a large impact on the long-term direction of EU climate policy.

In the lead up to The Hague negotiations, the Environment Council's emphasis on 'environmental integrity' reflected the view of many environment ministers and agency officials that the EU had conceded too much flexibility at Kyoto. As one environment official argued: 'we were not going to come home like we did from Kyoto and say to the EU "this was all we could get" again' (EU-23; EU-20; EU-32). In addition, some environment ministers, especially those from the Green Party, were antagonized by what they saw as a 'take it or leave it' approach from the US on key issues (EU-11; EU-19). The question for environment agencies was how to achieve a balance 'between flexibility mechanisms, on the one hand, and maintaining the integrity of the Kyoto protocol on the other' (EU-35). The reason, as a French official explained, was that 'there was a strong sense from environment ministers that if a commitment is made to a certain level of effort the effort should be made domestically' (EU-57).

In Europe, the discussion focused on two types of flexibility mechanisms: carbon sinks and emissions trading. The first was much less controversial internally than the second. For the majority of environment ministers and environment agencies carbon sinks were a totem for environmental integrity. They were determined not to broaden the definition of sinks, as the US wanted, because they argued that it would enable countries to avoid domestic emissions reductions (EU-19). The aim, as a former environment official put it, was 'to try and keep forestry credits out of the agreement as much as possible' (EU-20).

For the same reasons, environment ministers and agencies had also been suspicious of emissions trading believing some nations would try to 'buy their way out' of taking domestic action. However, after the failure of the carbon tax in the early 1990s, there was some recognition that new economic instruments would be needed if the EU was to meet its emissions target under the Kyoto Protocol (van Asselt, 2010). Germany was a case in point. The general position of the BMU 'was to push for a target and to get there' (EU-23). Under the EU burden sharing agreement negotiated in 1998, Germany was required to reduce its emissions by 21 per cent based on 1990 levels. Despite the reduction in emissions as a result of reunification, in March 2000, the Council of Environmental Advisors to the German Government announced that Germany would not be able to meet its emissions target without additional effort (Hatch, 2007: 52). As the international regime approach would expect, this triggered domestic realignments as state actors began to consider how they would meet international commitments under the Kyoto Protocol.

To varying degrees, environment agencies began to consider if emissions trading was the answer. The genesis of this inquiry began with the replacement of the 'true believers' by officials who were well-versed in economics. The key environment agency was DG Environment, which was able to play an entrepreneurial role from its position in the Commission. The personnel changes described above were crucial. As one interviewee pointed out, 'there was never going to be emissions trading under [Jorgen] Henningsen' (EU-36). This was not the case under his successor Jos Delbeke, who was now the Head of Unit responsible for climate change. He, along with his colleagues, including Peter Zapfel, had spent time studying emissions trading. And, as the transnational approach would expect, after Kyoto they maintained discussions with experts from the US, who had been actively involved in the Clean Air Amendments. On reflection, DG Environment officials and EPA officials who had participated in these discussions claimed that they played a crucial role in persuading the Commission that emissions trading was an idea worth pursuing (EU-46; US-22; EU–32; EU–23). As will be discussed in the following section, US environmental NGOs were also influential.

Within DG Environment, the so-called 'Bureaucrats for Emissions Trading (BEST) team', had to convince the Environment Commissioner, Ritt Bjerregaard, of the benefits of emissions trading (EU-49). According to a former member of her staff, she had taken a strong stance against the proposal in the lead up to Kyoto (EU-42). The strategy of the BEST team was to focus on the environmental certainty that an emissions trading system could provide by establishing an overall cap on emissions. Delbeke and his team argued that as long as the cap was set at the right level, the emissions reduced within the EU and reductions effectively monitored, emissions trading would work (Skjaerseth and Wettestad, 2008: Ch 4). It was an argument that proved persuasive and one that Margot Wallstrom, who replaced Ritt Bjerregaard as Environment Commissioner in September 1999, would also support (EU-19).

In addition, for many of these actors who had been involved in the failed carbon tax proposal, discussed in Chapter 3, the proposal for an emissions trading system had a political fillip that the carbon tax did not; qualified majority voting. Whereas the carbon tax, as a fiscal issue, required a unanimous vote to be carried, an emissions trading system only required a qualified majority of the Environment Council, which would make it much easier to pass. This was well-recognised by Delbeke and economists in DG Research and parts of DG EcoFin, who were now working with the BEST team to provide the economic modelling that would underpin an emissions trading proposal (EU-53; EU-60).

While a qualified majority might be easier to attain, there remained some skepticism among environment ministers and environment agencies about emissions trading. In Germany, the Environment Minister, Jurgen Trittin of the Green Party, 'was reticent about emissions trading from the beginning' (EU-26), as were some of his closest advisors within the BMU. The same was true of environment ministers and agencies in France and Belgium (EU-55; EU-19). However, even in these environment agencies there were economists working in concert with Delbeke's team. For example, Franzjosef Schafhausen, an economist and senior official in the BMU, was strongly in support of emissions trading (EU-32). Emissions trading was slowly becoming more acceptable, especially in agencies where the ministers were not from the Green Party. Another example was in the UK, which was one of the strongest proponents of trading, the Department of Environment, under Prescott, supported emissions trading (EU-35). In February 2000, Prescott and Prime Minister Tony Blair put their support behind a domestic emissions trading system (Skjaerseth and Wettestad, 2008: 88).

Then, in March 2000, DG Environment published a Green Paper on emissions trading, which marked a significant change in the position of the Commission (European Commission, 2000). The Green Paper was the culmination of earlier draft papers in 1998 and 1999. The most notable aspect was that none of the scenarios modelled assumed full international trading. The case was only made with reference to internal EU trading. This meant member states would have to rely on domestic emission reductions. This was consistent with the strategy Delbeke had first used to persuade the Environment Commissioner, Ritt Bjerregaard, that emissions trading would not provide loopholes for member states to avoid action. It was also consistent with the EU position to set a 50 per cent ceiling on the use of flexibility mechanisms. This would be necessary to ensure that countries outside the EU would also undertake the majority of their emission reductions domestically.[7] In other words, while the US imperative had been to have developing countries included in an international emissions trading scheme to keep costs down, the EU imperative was to keep them out, to ensure that the environmental integrity of the Kyoto Protocol would be maintained. While it would be another five years before an emissions trading scheme was established in Europe, there is some evidence here to suggest that the design being advocated influenced the type of agreement the EU was prepared to sign in The Hague.

Although environment agencies in Europe continued to lead the inter-agency discussions, as they had throughout the 1990s, they were not traditionally the most powerful state actors. Their influence on the EU position was directly related to the engagement of their counterweights inside government; namely the economic, energy and industry agencies. Many officials in the member states and the Commission argued that these agencies were not fully mobilised, especially among the senior ranks. For example, in the UK, one official claimed that the Treasury 'was not really engaged' and believed 'this was greeny stuff not core business' (EU-35; EU-36). Similar statements were made by officials in Germany and France (EU-33; EU-57).

However, as the implications of the Kyoto Protocol became apparent there was a small shift among these agencies, which is exactly the type of realignments that the international regime approach would anticipate. A case in point was the reaction of DG Industry to the Kyoto Protocol:

> Those negotiations at Kyoto had been a closed circle between DG Environment and the relevant ministries of environment. For those of us on the outside in economics and industry there was really no idea of what the consequences of the Protocol was. No-one in DG Industry had been part of the negotiations or part of the Kyoto Protocol. It was only after the Protocol that we began to think 'what are we going to do about this?' (EU-61).

While some officials began to shift their attention, the senior ranks never actively participated in the inter-agency discussions, nor did they produce the same level of opposition as their US counterparts. In many cases they were simply concerned with other policy priorities, or they were grappling with precisely what the Kyoto Protocol entailed. A case in point was emissions trading. For many agencies, especially economic agencies, which had traditionally turned to taxes for environmental problems, the concept of trading remained unfamiliar. For example, among officials in the BMWi in Germany there was much uncertainty:

> We were not very comfortable with it [emissions trading]. The main concern was that it had to be negotiated and implemented in a very short time and its impact on the economy was going to be unlike any other instrument and only a few experts understood it (EU-33).

Although similar views were expressed by other economic agencies, not all were opposed (Downie et al., 1998). In the UK, the Department of Treasury and the Department of Trade and Industry were both in broad support (Skjaerseth and Wettestad, 2008: Ch 4). Further, not all economic and industry officials were uncertain about the implications of emissions trading. As discussed, within the Commission some officials in DG EcoFin and DG Research were working in tandem with DG Environment in the development of the Green Paper on emissions trading.

Ultimately, progressive environment agencies supported by the Environment Council continued to dominate the inter-agency discussions, which shaped the EU position leading up to The Hague. The limited engagement of economic, industry and energy agencies, especially among the senior ranks, meant that in most cases there was never significant internal opposition to what environment officials and their ministers were proposing. While the picture that emerges is largely consistent with what the two-level framework would expect, on its own this perspective overlooks some factors behind the EU's behaviour. For example, the evidence presented in this section indicates that the network of government actors that emerged across the Atlantic, especially between DG environment officials and US officials had a large bearing on the genesis of emissions trading in the EU. This is exactly the type of policy coordination that the transnational approach would predict. Similarly, the domestic realignments among state actors that were triggered by the climate regime only become apparent when the insights of the regime perspective are considered. In short, each of the three perspectives helps to shed light on the behaviour of state actors in the EU.

Environmental NGOs have 'A Lot More Power'

In the lead up to The Hague, non-state actors continued to play a role as the EU grappled with the type of agreement it would sign. However, there were changes in the networks between state and non-state actors, which realigned after the Green Party moved into government in key member states. As COP 6 approached, these changes directly affected the influence of environmental NGOs and business groups.

Prominent environmental NGOs, such as EU-CAN, WWF and Greenpeace, remained part of a progressive coalition and together with the Environment Council their principal goal was to maintain the 'environmental integrity' of the EU's position (EU-20). This was reflected in their stance on the two main flexibility mechanisms: carbon sinks and emissions trading. For example, Greenpeace and WWF were against any broadening of the definition of sinks to include avoided deforestation. And, environmental NGOs maintained their opposition to emissions trading with the slogan 'trading pollution is not a solution', which they had used since Kyoto (Skjaerseth and Wettestad, 2008: 27). Though, as will be discussed, some NGOs changed their position as the negotiations approached (EU-58; EU-30; EU-55).

To make their presence felt, NGOs networked with government agencies, as they had throughout the 1990s. However, with five of the ministers on the Environment Council now from the Green Party, there was a shift in the networks among NGOs and state actors. Environmental NGOs developed much closer relationships with environment agencies. For example, in Germany, the Environment Minister Jurgen Trittin, and some of his staff in the BMU, such as Karsten Sach, had close ties with WWF (EU-30; EU-19). It was the same in France and Belgium where former NGO staff moved to work for environment ministers from the Green Party (EU-55; EU-19; EU-11; EU-20). In addition, Commission officials confirmed that DG Environment was continuing to consult NGO experts in preparation for COP 6 (EU-61). It is not hard to see why. According to one former NGO official, there were more people working on climate change in the European WWF office than the entire climate change unit in DG Environment (EU-58). As other researchers have shown, NGO expertise can be a useful tool to influence state actors, especially when the knowledge and expertise of state actors is limited, as it was in DG Environment (Corell and Betsill, 2008).

The shift in the networks between NGOs and state actors also enabled NGOs in some countries to engage more directly in 'accountability politics' (Keck and Sikkink, 1998). For environment ministers from the Green Party, environmental NGOs were considered their 'constituency' and this gave them more power to hold these ministers to account (EU-55). One former NGO official explained the dynamic:

> It gave the environmental NGOs a lot more power in the EU than they have now. The Green ministers would have to answer to us. Each morning it would be WWF or Greenpeace on the BBC or CNN. This might not have been important in other countries but in those five countries where you had these Green ministers who had been elected by these constituents and then you have the NGOs making them accountable (EU-30).

While NGO officials continued to devote their time and resources to the domestic processes to affect the EU's international position, as the two-level approach would expect, transnational networks also influenced the EU position. Emissions trading was a case in point. In January 1999, the Foundation for International Environmental Law and Development (FIELD) in the UK and the Center for Clean Air Policy (CCAP) in the US were commissioned by DG Environment to prepare design options for a European emissions trading system (Skjaerseth and Wettestad, 2008: 78).

Working alongside Jos Delbeke and the BEST team, NGO staff, such as Ned Helme, who had worked on the sulphur dioxide regime in the US, were 'very influential' in preparing the EU position for the negotiations (EU-23; EU-41; US-22). Interestingly, the strength of these transnational networks in informing EU thinking was also evident in the ripple effect this had on environmental NGOs in the EU, which had initially 'hated' emissions trading (EU-58). A former NGO official explained:

> I went to a meeting in Brussels in December 1998, Jos Delbeke asked about six or seven NGOs to a meeting. It was like Chatham House rules. He said to us listen DG Environment is focusing now on emissions trading. We said back to them are you guys crazy? He said hang on we do not have to have a US model of emissions trading, this is the EU. If we can get the cap right and all the emissions are reduced within the EU what is wrong with that? We said okay perhaps you're right. That was a key meeting for me … we changed our position after that. Greenpeace and Friends of the Earth took longer to change (EU-58).

Business Groups take a 'More Positive' Approach

In the late 1990s there were no dramatic changes in the compositions of the business community. As discussed in previous chapters, in the fossil fuel sector BP, Shell and industry associations such as EUROPIA remained the most prominent. Other broader industry groups were also engaged like the ICC and the more progressive World Business Council for Sustainable Development (Lacasta et al., 2007; Giorgetti, 1999).

In the aftermath of Kyoto, some of these business groups began to engage more constructively. For example, BP and Shell both took a 'more positive' approach (EU-49) with BP launching its new logo 'Beyond Petroleum' in July 2000 (Kolk and Levy, 2001). The ICC too, acknowledged that it 'had to regroup' after Kyoto and re-think its approach to the future (EU-39). As the regime approach would predict, the shift in engagement reflected the realisation within the business community that the discussions in Europe were turning to implementation. In other words, the climate regime was producing realignments among business groups. As one government official pointed out:

> Business groups became much more involved after Kyoto when they saw EU policy crystallizing around certain policies that may hurt their interests (EU-40).

Business groups, especially from the fossil fuel sector, 'worked closely' with energy and industry agencies. For example, a DG Energy official commented that their 'traditional stakeholders were the energy industry most often represented by European bodies like EUROFER' (EU-28). In DG Industry, business representatives were brought in-house for their expertise and 'employed on a part-time basis' to consult in preparation for The Hague (EU-61). While members of economic agencies said the views of business were taken into account, they did not have the same close relationships with non-state actors (EU-36).

As the EU discussions progressed, the position of business began to shift. Initially business groups, especially in the cement and steel industries, argued strongly for voluntary agreements with government (Skjaerseth and Wettestad, 2008: Ch 4). For example, a French official noted that 'business tried to convince the Department of Industry that everything could be achieved through voluntary measures' (EU-40). The same was true in Germany (EU-32). Yet as European governments moved toward mandatory measures business groups sought to intervene. The clearest example was emissions trading. In both the member states and the Commission, oil companies, in particular, successfully intervened to ensure that state actors moved toward trading rather than taxes. For example, by the end of 1998 both BP and Shell had announced plans to establish an internal emissions trading system. EUROPIA, a federation of 20 oil companies, which had actively opposed a carbon tax in the early 1990s also moved to support emissions trading (Skjaerseth and Wettestad, 2008: Ch 4; van Asselt, 2010; Christiansen and Wettestad, 2003; Meckling, 2011).

At the same time, state actors encouraged industry involvement. In the UK, which was the first country to introduce an emissions trading scheme, the Government had been 'very careful to involve business in its decision-making' and groups such as the CBI were 'influential' in the end design (Darkin, 2006: 270). Together these moves had a demonstrable effect on environment agencies that were working on the design of emissions trading schemes. A former BMU official in Germany explained.

> Another piece of the jigsaw was that it was very helpful to have BP explaining what they were already doing internally. They had an emissions trading scheme going internally, which had been successful at reducing emissions within the company. That was very persuasive for us and the environment ministry began to campaign to explain the BP scheme and how it works and its successes (EU-23).

That being said, it is difficult to determine which way the influence flowed. It is just as possible that the ultimate support from business also reflected the success of government officials, such as the BEST Team in DG Environment, demonstrating to industry the benefits of emissions trading. Indeed, some business representatives were more inclined to view the government outreach as an attempt to avoid the type of business opposition that had characterised the failed attempt to introduce a carbon tax in the early 1990s. Reflecting on the emissions trading discussions in the UK, one stated:

> God knows how many stakeholder meetings they had and they established a government business working group as well (EU-39).

Nevertheless, what is clear is that business groups in the EU became more engaged after Kyoto, and that their activities were almost exclusively conducted through national processes. While they did not have the same influence on the EU that US business groups had, their support for emissions trading was one example where they were able to shape the EU position leading up to The Hague negotiations. In sum, the evidence largely supports both the proposition of the regime perspective that the international regime will trigger domestic realignments, and the two-level proposition that COGs will monopolise the external representation of the state.

Why did the EU Not Sign an Agreement at The Hague?

At COP 6 in The Hague, the EU did not sign an agreement. The decision reflected a strong coalition of environmental interests supported by a COG whose preference was largely determined by environment ministers of the Green Party. The overriding objective of this coalition was to preserve the environmental integrity of the Kyoto Protocol. In many respects, little had changed since the Kyoto negotiations. The environment agencies remained largely unopposed by economic, energy or industry agencies. NGOs continued to support the position of environment agencies, and the limited engagement of business groups meant that unlike their counterparts in the US, they did not seek to act as a veto coalition. Once again, the two-level perspective best captures these factors and the effect they had on the behaviour of the EU. First, the preferences of the COG were critical. The electoral success of the Green Party and the resulting shift in the composition of the Environment Council ensured that it, together with the environment agencies in key member states, would pursue a strong climate agenda. To be sure, the historic dominance of the Green Party by November 2000 put environmental integrity at the top of the EU agenda.

Second, the shifting networks within the environmental coalition, most notably the replacement of 'true believers' by economists in some environment agencies, softened the approach that the EU took to emission trading. This was most evident in DG Environment where the growing influence of economists was linked closely to NGOs both in Europe and the US. It should be noted, that the role that US NGOs played in this respect also supports the proposition of the transnational approach that such networks can affect policy outcomes by enrolling powerful actors.

Third, the power of environmental interests resulted from the distribution of domestic coalitions as the two-level approach would predict. Despite a realisation by the economic, energy and industry agencies that in the wake of the Kyoto Protocol they would need to more actively engage in the inter-agency discussions, this had still not occurred among the senior ranks of these agencies by the time of The Hague. In addition, business groups who traditionally networked with these agencies were much more constructive in their approach (in marked contrast to US business groups).

These factors had a clear effect on the position that the EU took on several key issues. First, the strict position the EU maintained on carbon sinks was a manifestation of the preference of the Environment Council and its concern for the environmental integrity of the Kyoto Protocol. Without significant opposition from other agencies or non-state actors, there was near unanimous support for maintaining the definition of sink activities agreed to in Kyoto. The Chair of the Environment Council, French Environment Minister Dominique Voynet, epitomised the EU position. Referring to the US proposals on sink activities she stated:

> We do not intend to redebate this item. We want to discuss ways in which we can meet our commitments, not ways in which we can avoid them (cited in Brown, 2000: 15).

Second, the EU's position to accept emissions trading contingent on a 50 per cent ceiling reflected the ongoing internal debate in the EU. On the one hand, the decision to agree to emissions trading instead of binding policies and measures to reduce emissions, as it had at Kyoto, attests to the growing influence of economists within the environment agencies. Many of these economists, including the BEST Team, were strongly influenced by their interaction with government and non-government experts on the other side of the Atlantic. On the other hand, the decision to support a ceiling on the use of flexibility mechanisms reflected the political reality that these economists faced. As discussed, a ceiling was used to persuade Green environment ministers and the environmentally-minded Environment Commissioners, Ritt Bjerregaard and Margot Wallstrom, that flexibility mechanisms would not be exploited by member states to avoid domestic emission reductions.

Third, although the EU was open to voluntary commitments for developing countries, as it had been at Kyoto, it did not advocate for them in the same fashion as the US. To a large extent this revealed the dominance of the progressive coalition of environmental interests, which remained largely unopposed by other actors, such as industry agencies and business groups. It is likely that these actors would have voiced greater opposition had they been mobilised, but at this stage they were not. It is no surprise then that the Green Paper on emissions trading did not assume full international trading.

However, the decision by the EU not to sign an agreement was not simply because the agreement being mooted was different to the negotiating position it took to The Hague. After all, at both Berlin in 1995 and Kyoto in 1997, the EU had been prepared to sign an agreement that went beyond its initial position. The key difference this time was that the Environment Council, especially those ministers of the Green Party, believed that the EU had given away too much to the US to secure the Kyoto Protocol. The Hague was seen as an opportunity to redress the perceived imbalance:

> Basically, at Kyoto we did not push hard enough, we were too flexible, so the realisation at The Hague was that we had to be tougher on some issues. When we looked back at Kyoto the first reaction was, we had a deal, but then we looked at it further and we thought we had really been too flexible and we wanted to tighten up the Protocol (EU-23).[8]

In the end, as the two-level approach would anticipate, this resulted from the preferences of the Environment Council and their domestic political incentives. Whereas at Kyoto there was an imperative for many environment ministers to get an agreement, as we saw in the last chapter, by the time of The Hague the imperative was not to threaten the environmental integrity of the Kyoto Protocol. No agreement was now seen as better than a bad agreement.

CONCLUSION: ARE THE THEORETICAL PERSPECTIVES SUFFICIENT?

At the conclusion of the negotiations in The Hague in November 2000, the US and the EU were not able to forge an agreement. The two-level perspective best encapsulates the reasons behind the US and EU negotiating positions and their ultimate decision not to sign an agreement. In the US, a regressive coalition came to dominate domestic discussions, which when combined with the preferences of the COG, notably Al Gore's concern for his electoral prospects, ensured that the White House was cautious in searching for an agreement. In the EU, a strong progressive coalition combined with a COG dominated by the preferences of environment ministers of the Green Party, meant that the EU too was not willing to collude for the sake of agreement because their political imperative was to preserve the environmental integrity of the Kyoto Protocol. While there is also some evidence to support the propositions of the transnational perspective and the international regime perspective, ultimately; each of the three approaches is limited when applied to prolonged international negotiations. This becomes evident as we reflect on the changes in the behaviour of the US and the EU between COP 3 and COP 6.

In the three years following Kyoto, the US and the EU experienced significant shifts in domestic political dynamics, which led both parties in very different directions. In the US, the increased attention on the economic impacts of implementing the Kyoto Protocol intensified divisions within government. Within the executive, the economic agencies came to dominate the inter-agency discussions as the senior ranks became actively engaged. In the Congress, Republicans and powerful business groups re-framed the climate discussions along economic lines creating further pressure for the executive to address their demands.

In the EU, domestic politics was moving in a different direction. The historical high point of the Green Party in Europe meant that the principal concern for key member states, such as Germany and France, was environmental integrity, not economic impacts. While economic implications were becoming more important, especially as economists took on key positions in environment agencies, they had not yet reached the senior ranks of key agencies. Economic, energy and industry agencies were only beginning to engage in the inter-agency discussions and many officials were still grappling with the likely implications of the Kyoto Protocol. As a result, economic concerns around implementation remained subordinate to the overriding objective of the Environment Council; environmental integrity.

As discussed in the previous chapters, the temporal dimension of long negotiations is not well captured by the three theoretical approaches. Once this is taken into account it becomes clear that state preferences are fluid not fixed. Indeed, several factors appear to be critical for explaining the changes in the behaviour of the US and the EU and yet they lie outside the approaches employed here. First, is the level of engagement or mobilisation of actors; in prolonged international negotiations the engagement of state and non-state actors does not remain the same over time. For example, in the US, the senior ranks of the economic agencies, which had only started to mobilise prior to Kyoto, were completely engaged by 2000. In fact, the assumptions that allowed the economic modelling conducted by the CEA to conclude that the US could meet its Kyoto commitments at a 'relatively modest' cost, largely determined the US position on developing country commitments and flexibility mechanisms. In the EU too, as the implications of the Kyoto Protocol became apparent, there was a small shift among the economic, energy and industry agencies, though not yet at the senior levels. In addition, European business groups became more involved as 'they saw EU policy crystallizing around certain policies that may hurt their interests', such as emissions trading (EU-40). The regime approach, which proposes that the international climate regime will trigger realignments among domestic actors, captures some of these influences, though not all.

Second, the mobilisation of new actors can potentially affect the networks of state and non-state actors. For example, in the EU environment agencies under the leadership of environment ministers from the Green Party often had a closer relationship with NGOs than they had prior to Kyoto. This was the case in Germany. Another example was the strengthening transnational ties between officials in DG Environment with US officials in the EPA, and with American NGOs such as the Center for Clean Air Policy. These shifting networks are crucial to explaining the EU's about-turn on emissions trading after Kyoto.

Finally, while the two-level perspective points to the preferences of COGs for explaining state behaviour, it does not sufficiently account for changes in COG preferences that result from a change in their beliefs or political incentives. For example, in the US and the EU new domestic dynamics meant new political imperatives. For the White House, the success of Congress and business groups in re-framing the domestic debate along economic lines meant that any outcome that did not meet the demands of these groups would be politically explosive for the administration. Further, having delivered to the environment constituency by signing the Kyoto Protocol, it is likely the White House believed that no agreement would be better than a bad agreement. For the EU Environment Council the changes were different but the ultimate imperative was the same. With the dominant ministers from the Green Party believing that the EU had conceded too much ground at the Kyoto negotiations, there was strong pressure, especially from environmental NGOs, to deliver an outcome that was seen to maintain the environmental integrity of the Protocol. NGOs such as WWF and Greenpeace, who were a key part of their constituency, were ready to hold them to account if they failed. This meant that the EU also viewed no agreement as better than a bad agreement, though for very different reasons.

NOTES

1. As another administration official pointed out, it was not just Republicans who were opposed: 'The relationship with Congress in those years was very difficult both with Democrats and Republicans. There were some supportive Democrats but many were against Midwestern Democrats, people like Senator Byrd, were very concerned about coal, Senator Dingle on the auto industry. They were two key senators that were very opposed' (US-19).
2. A former State Department official reflected on the diplomatic strategy: 'In retrospect this was totally ridiculous. Each country had deep-seated national positions and we were sending diplomats all around the world to say the same things. The strategy was that in by-passing negotiators and appealing to ministers and heads of states at the upper levels we would be able to roll these deep-seated positions from the top down. In retrospect we were naïvely optimistic because we had some early wins, for example with [President] Carlos Menem in Argentina' (US-34).
3. In addition, according to one official, though it could not be corroborated, Treasury was demanding that the CEA not employ economic models that show minimal cost implications because this would increase the likelihood of the US taking action (US-23).
4. One official noted that the relationship between the White House and Treasury was often strained because the Vice-President had doubts about Lawrence Summers ever since his infamous memo while Chief Economist at the World Bank, which claimed that developing countries were under-polluted (US-23).
5. While the Kyoto Protocol required the EU to reduce its greenhouse gas emissions by 8 per cent on 1990 levels for the period 2008–12, the burden sharing agreement agreed to in the EU in 1998 meant that some countries could increase their emissions while others reduced them. This enabled the EU to placate countries such as Spain and Portugal, who were able to increase their emissions by 15 per cent and 27 per cent respectively, while other countries like Germany and the UK were required to decrease their emissions by 21 per cent and 12.5 per cent respectively.
6. Where there is no consensus in the Environment Council, a decision is reached by qualified majority voting, a system in which larger states have more votes than smaller ones. At the time of The Hague, the UK, France, Germany and Italy had the most votes with 10 each; followed by Spain with 8; Portugal, the Netherlands, Greece and Belgium with 5 votes each; Sweden and Austria with 4; Denmark, Finland and Ireland

with 3; and Luxembourg the smallest state had two votes. In this system a winning coalition needs 62 of the 87 votes while a blocking coalition needs 26.

7. The importance of these restrictions was also evident in the resolutions of the European Parliament. In October 2000, a resolution drafted by the Environment Committee, which largely supported the EU negotiating position was adopted by 437 votes to 6. It called on the EU to take the lead in the climate change negotiations and explicitly supported the ultimate position the Council took on sink activities and on a ceiling for the use of flexibility mechanisms (European Report, 2000b). While some members of the Environment Committee had viewed emissions trading as immoral when it was debated in committee in September 1999, by 2000 most MEPs were in support. However, this support was conditional on a ceiling (Skjaerseth and Wettestad, 2008: 80–81).
8. To compound this, The Hague negotiations were taking place during the Florida re-count of the 2000 presidential election. As a result, the Environment Council was uncertain if concessions made would be honoured by the incoming administration. As one former EU negotiator argued, ‘it was clearly not worthwhile having an agreement with … an American administration that was a lame duck at that stage’ (EU-26).

PART III

Empirical findings and theoretical implications

6. Discussion: the behaviour of the US and the EU in the international climate change negotiations

The first international negotiation I attended was during 2008 in Bonn, sparking my interest in what lay behind the behaviour of states. Why did they sometimes choose to cooperate and at other times choose to obstruct progress? As I showed in Chapter 2, many scholars have explored such questions and developed theories to explain state behaviour. For example, the two-level intergovernmental perspective, developed by Robert Putnam, relaxed the assumption of the unitary state to emphasize the role that state and non-state actors operating within national processes have on government decisions. In contrast, the transnational perspective focussed on the cross-boundary activities of these actors, whereas the international regime perspective highlighted the pathways by which regimes can affect the behaviour of states.

However, as I began to look across the years and decades of the international climate negotiations I was puzzled by the variations in state behaviour, like those of the US and the EU. As I have discussed in the preceding chapters, during the Kyoto phase of the negotiations the positions of both these parties, and the type of agreement that they were prepared to sign, changed. In 1995 the US and EU agreed to the Berlin Mandate, which stipulated no binding emissions targets and timetables for developed countries, no new commitments for developing countries and no flexibility mechanisms. Then in 1997, the US and the EU agreed to the Kyoto Protocol, which included binding emissions targets and timetables for developed countries and flexibility mechanisms. Finally, by 2000 the US and the EU refused to sign an agreement, which would have fleshed out the detail of the Kyoto Protocol already agreed to in 1997.

In this book I have set out to explore the puzzling variations in the behaviour of the US and the EU in the climate negotiations. What factors lead a state to change its position and the type of agreement it is willing to sign? And, how and why are these decisions made? In this chapter I will answer these questions by drawing together the data presented in the preceding chapters to critically analyse the propositions derived from each of the theoretical perspectives identified above, beginning with the two-level framework.

In the final chapter to follow, I will take this one step further by sketching an 'ideal type' framework for understanding the temporal dimension of prolonged international negotiations and prescribing specific strategies that state and non-state actors can use to shape international outcomes. After all, most of us, whether we are academics, government officials, members of NGOs, business groups, or students of negotiations seek to understand negotiations so we can steer them to our preferred outcome.

THE TWO-LEVEL INTERGOVERNMENTAL PERSPECTIVE: A DOMESTIC AND INTERNATIONAL GAME?

The two-level perspective proposes that in international negotiations the COGs monopolize the external representation of the state. While states largely respond to domestic pressures when forming their positions, especially the preferences of domestic actors and the distribution of domestic coalitions, the unique position of the COGs mean that they have some capacity to manipulate domestic pressures to their own preferences. These dynamics are also affected by the type of domestic political institutions and the process for ratification.

The previous chapters show that there is good support for these propositions in the cases of the US and the EU. First, while the COGs did not monopolize the external representation of the state, they certainly dominated. Sub-units of government and non-state actors did participate in the international negotiations, but as I will discuss in the next section, their role and influence was limited and, more often than not, when they did participate it was primarily under the terms set out by the COGs.

In the US, the White House, as the COG, coordinated discussions at the international table. The principal representatives from the State Department, such as Timothy Wirth, Stuart Eizenstat and Frank Loy, who led the US delegation to the international negotiations, all operated under the imprimatur of the White House, most notably of Vice-President Al Gore. For example, when Timothy Wirth announced in Geneva in 1996 that the US would support a legally binding agreement, he did so with the consent of the White House. When EPA or DoE officials participated in the international arena, it was almost always under terms prescribed by the White House.

The same was true in the EU, where the Council of Environment Ministers, with the mandate to negotiate, acted as the COG and dominated the external representation of the EU. EU environment ministers and their environment agencies coordinated the international discussions and, for the most part, it was the environment ministers who set the course. This was the case in Berlin, Kyoto and The Hague. Indeed, it was at The Hague in 2000, that the terms of the EU's participation were determined by the Environment Council's insistence that the 'environmental integrity' of the Kyoto Protocol be preserved, notwithstanding the increased participation of other state and non-state actors.

Second, and crucially, Putnam argued that any two-level theory must be rooted in a theory of domestic politics because the power and preferences of domestic actors will shape the level II win-set, that is, the set of possible international agreements that would have the necessary support to be ratified domestically (Putnam, 1988: 442). Accordingly, the two-level perspective proposes that a state's position is formed largely in response to the distribution of domestic coalitions. As will be apparent from the preceding chapters, this proposition is well supported by the available evidence.

On the surface, two broad coalitions can be identified in the US and the EU; a progressive coalition largely in support of climate change action and a regressive coalition largely opposed to it. Of course, below the surface the reality was more complicated with each coalition comprised of a series of networks of state and non-state actors not always in agreement and employing different strategies depending on the political environment. Hence, as the negotiations evolved so too did the power and preferences of these coalitions.

In the US, negotiations at Berlin in 1995 and Kyoto in 1997 were largely dominated by the preferences of the progressive coalition. Within the executive, a bureaucratic coalition, comprising the White House, the EPA and the State Department, which one member of the State Department labelled the 'White Hats', emerged to have a disproportionate influence on the type of agreements the US were willing to sign (US-25). Senior figures in these agencies, such as Timothy Wirth and Eileen Claussen, played crucial roles in navigating the inter-agency process to ensure that the US would agree to the Berlin Mandate in 1995, and significantly, would support a legally binding agreement, as announced a year later in Geneva, despite the potential opposition from other coalitions within and outside government. The White Hats were supported by a coalition of environmental NGOs including EDF, NRDC and the

Sierra Club. Personal networks between NGOs and members of the White Hats ensured that they were tightly networked to government agencies. One former official quipped that 'there were no real NGOs in the Clinton years … because they were all working in government' (US-14). When these NGOs were able to exploit these networks, the effect was to directly influence the preferences of the progressive coalition. A case in point was the role of EDF in the lead up to Kyoto. As discussed in Chapter 4, EDF staff were able to infiltrate government policy networks to ensure that they were 'in the room' when decisions were made about the US position on emissions trading.

A similar progressive coalition existed in the EU. It comprised environment agencies in the member states and the Commission. For example, in the lead up to Kyoto, the environment agencies in the UK, Germany and the Netherlands, among others, coordinated their actions to shape the EU negotiating position. At the same time, a related network of 'true believers' based in the Commission, which included the Environment Commissioners, and key officials in the member states worked to ensure that the EU would support an ambitious emissions reduction target at Kyoto. Part of the success of this group, at least in the short-term, stemmed from their strategy to keep 'climate change off the radar screen of other people' in order to minimize likely opposition (EU-49). After Kyoto, a new network emerged within the progressive coalition, the so-called 'Bureaucrats for Emissions Trading (BEST) team', which together with officials in other environment agencies, played a critical role in moving the EU toward accepting an agreement that included flexibility mechanisms such as emissions trading. And, in a similar fashion to the US, NGO networks within this coalition were also successful at infiltrating EU policy networks, as WWF showed with the development of the Triptique approach to burden sharing, which enabled the EU to take a 15 per cent emissions reduction target to the negotiations in Kyoto.

A second, more regressive coalition also existed, which was largely opposed to action on climate change. Whereas the progressive coalition worked concertedly throughout the 1990s to push the US and the EU to sign ambitious agreements, the regressive coalition, namely powerful sections of the business community often in concert with networks of economic and energy agencies, intervened as a veto coalition. These interventions were typically in reaction to proposals that would likely disadvantage the interests of these actors, as the two-level perspective would anticipate. Interestingly, the engagement of this coalition increased as the negotiations wore on, most noticeably as discussions turned to ratification of the Kyoto Protocol.

While this coalition, led by business groups, did intervene to successfully veto the BTU tax in the US in 1993 and the carbon tax in the EU in 1994, for the most part these actors were not engaged. However, by 1997 this coalition began to mobilize as the international agreements proposed tangible implementation requirements. For example, in the US, coal companies such as Peabody Coal and American Electric Power together with the GCC, were successful in initiating the Byrd-Hagel Resolution, which was passed unanimously by the US Senate just months before Kyoto. By articulating the economic concerns of industry groups, the Resolution directly constrained the US negotiating position both in Kyoto in 1997 and in The Hague in 2000. While EU business groups were less engaged as a veto coalition, like their counterparts in the US, their activities were largely confined to the national not the transnational level.

Further, the reason that US business groups focussed their attention on the Senate, whereas EU business groups, to the extent that they were engaged, targeted electoral politics instead of the legislature, reflects the different type of domestic political institutions, as the two-level perspective would assume. In the US majoritarian political system, domestic groups devote their attention to pressure points of ratification like the US Senate where a two-thirds majority is required. In contrast, in proportional electoral systems, such as those in the EU, interest groups instead focus on election campaigns given that ratification is typically a fait accompli once an agreement has been signed (Harrison and Sundstrom, 2007; Kroll and Shogren, 2008).

The two-level perspective implies that the relative homogeneity of coalition interests and the type of strategies they employ will affect their capacity to shape the domestic win-set (Evans, 1993). Both of these factors are evident in the case of the Byrd-Hagel resolution. First, in 1997 when the Resolution was passed, the costs of any agreement were likely to be borne by a concentrated group of fossil fuel producers who were highly organized under umbrella organizations like the GCC. The homogeneity of their interests and their high levels of organization meant that they were likely to be a very effective veto coalition. Second, the influence of the Byrd-Hagel resolution, as administration officials conceded, reflected its success in framing the public debate. As noted in Chapter 4, there is much empirical evidence to suggest that actors who strategically frame an issue that is consistent with their goals are most likely to reap the rewards.

The regressive coalition also comprised economic, energy and industry bureaucracies, which provided a critical counterweight to the preferences of progressive government agencies. This was most evident in the US. For example, in 1998 as the discussion focussed on the economic impacts of the Kyoto Protocol, economic agencies began to forcefully push the case that mitigation would require 'heavy economic lifting', in opposition to what the EPA and others claimed. As the salience of the issue increased, the interests of agencies such as the Department of Treasury and the CEA were fully incorporated into the US negotiating position. The influence of these state actors was less evident in the EU. For example, in the UK the Department of Treasury and the Department of Trade and Industry 'were more or less aligned against the Department of Environment'. However, even by the late 1990s most of these agencies were 'not really engaged' and, as a result, the preferences of the progressive bureaucratic coalitions in the EU continued to dominate. The hand of the progressive coalition was also strengthened by the fact that the domestic political environment was more supportive with strong public backing for action on climate change in nations such as Germany and the Netherlands.

The limited engagement of other groups that would likely oppose action on climate change, especially in the first half of the 1990s, largely explains why the preferences of the progressive coalition shaped the positions of the US and the EU. After all, environmental interests traditionally have limited power in government deliberations, yet because other actors such as economic agencies and business groups were not mobilized, as one environment official commented, 'you could get away with doing things that you could never get away with if everyone was focussed on it' (EU-37). This, in turn, reflected the low domestic salience of the issue, especially in the US, and the fact that prior to Kyoto, negotiations remained in a bargaining phase, ratification with tangible implementation requirements was not yet being discussed.

A third proposition of the two-level intergovernmental perspective is that the COGs are able to manipulate domestic pressures. However, the strategies a statesman employs will be dependent on their preferences, which may not always be the same as those of his or her constituents. In the US and the EU, the COGs held strong preferences about the type of agreement they preferred to sign, informed both by personal beliefs and the desire to enhance their domestic political positions. In the US, the Clinton White House was determined to 'distinguish itself' from the previous Bush administration on the issue of climate change. Vice-President Al Gore had campaigned strongly on the issue both as a Senator and as a Vice-Presidential candidate and he brought with him former advisors who had publicly stated their desire for action on climate change. In the EU too, environment ministers who sat on the Environment Council, such as Angela Merkel and John Prescott, were committed to climate action. Further, as we saw in the lead up to The Hague, the historic dominance of environment ministers from the Green Party increased this desire. In both cases there is evidence to suggest that the release of IPCC reports and interactions between COGs and national scientists also fortified the preferences of national leaders. Senior officials in European environment agencies who worked alongside ministers on the Environment Council claimed that they tried to push for positions 'that were as close as possible to what the science was calling for' (EU-32; EU-19; EU-26).

The preferences of COGs also reflected political imperatives as they worked to enhance their domestic political position. In some instances this was translated into a strong preference for agreement. For example, in the US an agreement at Berlin in 1995 and at Kyoto in 1997 was part of the White House's domestic political strategy to deliver to its environmental constituencies. As one former White House official commented, 'we really wanted to do anything to get an agreement' (US-6). Yet by 2000, concerned about his electoral prospects, the Vice-President and his advisors wanted the White House to downplay the issue and as a result there was not the same determination to reach an agreement in The Hague. Likewise, in 1995, the German Government under Chancellor Kohl and Environment Minister Angela Merkel made it clear to the world that an agreement in Berlin was a priority for the government, as it was for other western European leaders. The political imperative was equally as strong for environment ministers from the Green Party, such Dominique Voynet in France. As one former advisor stated in relation to the Kyoto conference, 'if there was no agreement at Kyoto it would have been a political failure for her' (EU-47). Indeed, at Berlin and Kyoto there is good evidence to suggest that one of the main reasons an agreement was reached was because the COGs in the US and the EU were happy to collude and offer concessions in order to secure an agreement, which was their primary political goal, whereas this was clearly not the case at The Hague three years later.

To achieve outcomes close to their preferences, COGs attempt to manipulate domestic constraints such as the power and preferences of domestic coalitions. A case in point was White House interventions to manipulate bureaucratic coalitions. Part of the reason the State Department had a relatively free rein in the domestic discussions in the early 1990s and Timothy Wirth, the lead US negotiator at the time, was able to agree to the Berlin Mandate, was because, as one former State Department official pointed out, the 'Vice-President was the guy who bulletproofed them from the rest of the departments' (US-9). Again at Kyoto, it was the Vice-President who intervened and said 'take the agreement' despite the opposition from economic agencies (US-2). In addition, the

White House also attempted to use side-payments at the international level to bring developing countries, such as Argentina and Kazakhstan, into any international agreement in order to appease those parts of the regressive coalition that were threatening to veto ratification of the Kyoto Protocol. There is less direct evidence of this manipulation in the EU, though it is clear from interviews with members of the progressive coalition, especially in the environment agencies, that the implicit support they had from environment ministers and heads of state, for example, in Germany and the UK, provided them with the necessary autonomy to develop negotiating positions in line with the preferences of their statesmen. Of course, in the EU the comparative lack of opposition from other actors, both inside and outside government reduced the need for manipulating the power and preferences of domestic coalitions.

In sum, the two-level intergovernmental perspective highlights many of the most important factors behind the behaviour of the US and the EU and their willingness to sign international climate agreements. In particular, it draws our attention to how the US and the EU responded to the power and preferences of domestic coalitions and their associated networks when forming negotiating positions. In addition, it draws our attention to how and why COGs can and do manipulate such domestic constraints. In other words, this framework captures most of what took place in the case studies. However, as will become clear in the following sections, without the insights of the transnational and the international regime perspectives, our understanding of the US and EU behaviour would be incomplete. Further, and crucially, this perspective does not appreciate the temporal dimension of prolonged international negotiations, such as changes in the level of engagement of actors or the distribution of domestic coalitions. This is an issue I will return to in the final chapter.

THE TRANSNATIONAL PERSPECTIVE: ARE STATES STILL IN CONTROL?

The transnational perspective differs from the two-level perspective because it proposes that the cross-boundary activities of actors matter in world politics. In other words, the COG does not monopolize the external representation of the state. Therefore sub-units of government, or bureaucracies, and non-state actors that engage directly with each other through transnational networks have the capacity to affect state behaviour by influencing domestic and international policy outcomes. These networks can include transgovernmental networks, transnational advocacy networks and epistemic communities, among others.

While the evidence from the case studies does not support the propositions of some theorists of a 'new world order' of transgovernmental networks (Slaughter, 2004), or the rise of a 'world civic politics' (Wapner, 1995), it does show that statesmen no longer have a monopoly on the external representation of the state. The most widespread example is the transnational advocacy networks of environmental NGOs, which are ubiquitous in the climate change arena. NGOs such as Greenpeace, EDF and WWF coordinated their participation in the international negotiations under the umbrella of the Climate Action Network. Business groups too, though in a more limited way, engage in activities across national boundaries as do sub-units of government, which participate in transatlantic transgovernmental networks.

Consequently, these actors have opportunities to manipulate domestic pressures and trigger realignments in the same way that the two-level perspective assumes only statesmen can. There is some evidence to suggest that transnational activities influenced US and EU behaviour. First, transgovernmental networks engaged in policy coordination and coalition building. In the early 1990s EPA officials and European environment officials participated in a series of policy dialogues around market-based mechanisms to reduce emissions. This provided a critical opportunity for European officials to learn about emissions trading and the US experience with the Clean Air Act Amendments. By the mid-to-late 1990s, actors within this network had built coalitions between economists within the European Commission, who had established the so-called Bureaucrats for Emissions Trading (BEST) network, and EPA officials who had convinced the BEST network that emissions trading was a concept worth pursuing. However, as the transnational perspective would anticipate, transgovernmental networks is too narrow a concept because it is limited to state actors. This was certainly true in the case of emissions trading where European bureaucrats also relied on the input of US NGOs, such as the Center for Clean Air Policy, in the lead up to COP 6 in 2000.

Indeed, the cases of the US and the EU suggest that broader transnational networks that include non-state actors are more prevalent and equally or more effective at influencing domestic and international policy outcomes. For example, the transnational networks of environmental NGOs employed information strategies with significant effect. As discussed in Chapter 3, EU negotiators acknowledged that NGO networks had worked 'as a back channel helping you to know what other governments think' (EU-37). In addition, the Green Group of progressive developing countries and the EU, which was formed in the course of the negotiations in 1995, and helped to make the Berlin Mandate possible, was a result of interventions by an NGO network led by WWF.

Business groups also saw the potential to engage in transnational activities that could trigger realignments. A case in point was the Climate Council, a US lobby group, which formed alliances with OPEC countries, notably Saudi Arabia, to pressure the US and other countries into a more restricted agreement. However, there is limited empirical evidence in the preceding chapters to suggest that these transnational networks directly influenced the positions of the US and the EU, or what they were prepared to sign. This was especially so for business groups, who worked mainly at the national level to secure their principal objectives.

The transnational perspective also proposes that epistemic communities, like the IPCC, will promote organizational learning by helping to create shared understanding based on their specialized knowledge. There is evidence in both the US and the EU to support this proposition, though it is notoriously difficult to ascertain the precise influence of the IPCC on individual agreements. That being said, US government officials, for example, were quick to point to the impact that the release of the IPCC Second Assessment Report in December 1995 had on domestic discussions. As one interviewee recalled, after the release 'the whole issue got elevated to a deputy secretary level where the ticket to entry was that everyone had to have a bricfing on thc science' (US-7). The increased penetration of climate science into the circle of policymakers affected the preferences of statesman like US Vice-President Al Gore and many of his advisors. The scientific arguments were also employed by members of the progressive coalition to help justify their positions, such as the declaration in Geneva in 1996 that the US would support binding commitment. Likewise, interviews with EU officials indicate that members of the 'true believers', for instance, were bound by a common belief in the science informed by the IPCC, and many European heads of state, such as Tony Blair and Helmut Kohl made public statements to the same effect.

In summary, the transnational perspective is useful to the extent that it highlights the cross-boundary activities of state and non-state actors, which are not captured by the two-level approach. Despite this, there are only limited instances where these activities seemed to have had an impact on the type of agreement the US and the EU were prepared to sign. Instead, it is the actions of state and non-state actors through domestic processes, as seen above, which have had the most influence. In short, the US and EU COGs remained the primary actors in the international realm and there is little evidence to suggest, at least in the cases examined here, that the prevalence of transnational networks has been translated into equivalent influence.

THE INTERNATIONAL REGIME PERSPECTIVE: DO INTERNATIONAL REGIMES MATTER?

The international regime perspective proposes that international regimes are an important part of world politics because of their capacity to influence the behaviour of state and non-state actors and the networks in which they operate. As a result, this perspective overlaps with the previous two in so far as international regimes help to constitute two-level games with basic rules and norms, and help to constitute the environment in which transnational networks operate. This perspective then, proposes that regimes influence actor behaviour through utilitarian and non-utilitarian pathways. These include: affecting the cost-benefit calculations of actors; acting as learning facilitators, which gives rise to social learning; and being an agent of internal realignments by shifting the balance of power among interest groups.

As international regime scholars acknowledge, critically analysing these propositions by specifying the ways in which regimes make a difference to actor behaviour is fraught with difficulty (Breitmeier et al., 2006). Nevertheless, there is some evidence from the cases of the US and the EU to support these propositions and suggest that the international climate regime has affected state behaviour. First, the international climate regime seems to have influenced the cost-benefit calculations of COGs by manipulating intangible costs, namely reputation. In the lead up to Berlin in 1995, the Clinton administration signalled a strong desire to 'distinguish itself' from the Bush administration and restore the US international reputation on climate change. Reputational costs were also on the minds of US negotiators in 1997. Months before the Kyoto conference was to commence, the US was ridiculed by the international community for announcing it would not sign an international treaty to ban land mines negotiated in Oslo. According to one senior negotiator, the instructions prior to Kyoto were 'don't let this be Oslo' (US-9). There were similar pressures on some member states in the EU. For example, former British officials admitted that in the early 1990s 'there was a very strong desire for the UK to be seen as strong on environmental issues' because a series of environment ministers in the Conservative government wanted to cast off the reputation the UK had as the 'dirty man' of Europe (EU-45).

Second, the international climate regime affected state behaviour through non-utilitarian pathways as well. One example is the path by which European bureaucrats learnt about emissions trading. As discussed, transnational networks played an important role in enabling policy coordination across the Atlantic. However, the transnational perspective does not explain why such networks were created in the first place. The international regime framework proposes that it is regimes that initiate such processes and, therefore, act as learning facilitators. This would seem to be the case here. Without the US and the EU actively looking for mechanisms to reduce emissions because of their obligations, or likely future obligations, under the climate regime, it is much less certain that such transatlantic policy networks would have been established. The international regime and the associated negotiations around legally binding emission targets encouraged state and non-state actors to learn from the experiences of their counterparts, as the EU did from the US experience with the Clean Air Act Amendments. In addition, there is some evidence to suggest that the international ozone regime also provided learning opportunities about emissions trading for EPA officials, which in time were grafted onto domestic and international climate discussions.

The international climate regime also triggered realignments among state and non-state actors in the domestic sphere, which influenced US and EU behaviour. As the climate regime began to detail legally binding commitments under the Kyoto Protocol, it had the effect of mobilizing different networks of actors. For example, US economic agencies began to play a more active role in the inter-agency discussions in the late 1990s for this reason. Likewise, business groups, principally fossil fuel producers, did so too because they believed that obligations under Kyoto, if met, would negatively affect their interests. While the effects were not as great in the EU, officials from economic and industry agencies did note that 'it was only after the Protocol that we began to think what are we going to do about this' (EU-61). At the same time, the international regime helped to institutionalize the role of climate scientists within inter-agency discussions and facilitate interactions between members of the IPCC and policymakers. Further, the development of the climate regime helped to empower and legitimize the activities of environmental NGOs. Working through national and transnational networks, NGOs were able to name and shame states that walked away from their international commitments under the regime. In the words of the transnational perspective, it enabled NGOs to more effectively engage in 'accountability politics', that is, the effort to hold powerful actors to their previously stated policies or principles (Keck and Sikkink, 1998).

In sum, there is evidence to suggest that the international climate regime influenced the behaviour of the US and the EU via specific behavioural pathways. However, with only two case studies it is difficult to isolate how much of the variance in behaviour of the US and the EU can be attributed to the international climate regime. This is especially so given the overlap between the three theoretical perspectives. That being said, excluding the propositions of international regime theory may provide greater parsimony, but it would also limit some important insights which contribute to understanding the type of agreements the US and the EU were willing to sign. Given the evidence presented here it seems difficult to assert that the international climate regime did not matter. The question that remains for future research is; how much did it matter?

CONCLUSION: THEORETICAL LESSONS AND LIMITS

Each of the three theoretical perspectives analysed in this chapter are valuable because they highlight the role of different actors – state and non-state – at different levels of analysis – domestic, international and transnational. The two-level perspective demonstrates the importance of the power and preferences of domestic coalitions and networks to shaping US and EU behaviour, and the role COGs have in manipulating such domestic pressures. The transnational framework illustrates that the two-level approach is too narrow because it ignores cross-boundary activities that state and non-state actors pursue to influence state behaviour. Finally, the international regime perspective points to the role that regimes play by changing the cost-benefit calculations of states, initiating learning processes, or triggering domestic realignments.

However, in light of the evidence, it is the two-level intergovernmental perspective that most convincingly explains the negotiating position and the type of agreement both parties were willing to sign. It explains why the US and the EU signed the Berlin Mandate, pointing to the dominance of progressive coalitions in both cases, supported by willing COGs, whose preference for agreement ensured that they would collude as necessary to produce an agreement. It explains the outcome at Kyoto for very similar reasons. The progressive coalition was able to dominate, despite opposition from some state and non-state members of the regressive coalition. The preferences of the COGs, informed by the science and their own domestic political imperatives, intervened as necessary both at the domestic and international levels to secure the Kyoto Protocol. Finally, this perspective also explains why negotiations broke down at The Hague in 2000. In the US the regressive coalition came to dominate domestic discussion inside and outside government. These domestic constraints combined with the preferences of the COG, notably Al Gore's concern for his electoral prospects, meant that the White House was less willing to search for an agreement. In the EU, a stronger progressive coalition combined with a COG dominated by the preferences of environment ministers of the Green Party, meant that the EU was not willing to collude for the sake of agreement because their political imperative was ensuring the environmental integrity of the Kyoto Protocol.

While these are the most critical factors, as seen above, they do not provide a complete explanation of how the relevant negotiations played out. They do not capture, for example, the important intervention of transnational networks of NGOs at Berlin, which helped to bridge the divide between parties to secure agreement. Or, for example, the influence of epistemic communities, namely the IPCC, in informing the preferences of the COGs and working to shift the domestic political dynamics. Or, for that matter, the role of transnational networks around emissions trading, which, facilitated by the international climate regime, helped to move the US and the EU positions closer together on emissions trading. Rather it is the propositions of the transnational perspective or the international regime perspective that provide more convincing explanations for these episodes. Nevertheless, taken overall, the evidence gathered in the preceding chapters provides only limited support for each of these theories.

However, and this is critical, the analysis so far only captures the factors behind the US and the EU behaviour in each set of negotiations. It provides a good account of how and why the US signed the Berlin Mandate, the Kyoto Protocol and did not sign an agreement in The Hague. Yet it does not sufficiently address the central puzzle of this thesis; that in prolonged international negotiations the type of agreement a state is prepared to sign can change over time. In particular, it does not address what factors led the US and the EU to change their positions? And, how and why these decisions were made? In other words, it fails to account for the temporal dimension of long negotiations. That said, if the analysis of each of the three negotiations is considered together it is possible to address this puzzle by identifying the critical factors behind the US and EU behaviour and then tracing how, if at all, they varied from 1995, to 1997, to 2000. In the final chapter I take up this task in order to improve our understanding of state behaviour in prolonged international negotiations.

7. Toward an understanding of prolonged international negotiations

The international climate change negotiations represent a perfect illustration of prolonged international negotiations. However, as discussed in the introduction, they are no ordinary set of negotiations. The problem they are trying to address threatens the very survival of our species, and for that matter, many others, on this planet. This is the consensus of the international scientific community. It is also the reality that many of the actors in this book grapple with as they negotiate the latest round of international climate negotiations. Many lament the failure of these negotiations, and it is true, global greenhouse gas emissions continue to spiral out of control despite the best efforts of previous negotiations. Yet they have also produced some remarkable successes, such as the Kyoto Protocol in 1997, which many believed was impossible only years before.

This book has been about the successes and the failures, but most importantly, it is about the lessons they hold for those who want to make a difference in future prolonged international negotiations be they on climate change, trade or security issues. This is the focus of this chapter, the lessons, and more particularly, the strategies and variables that influence states in prolonged international negotiations. The lessons may be drawn from history, but the theoretical insights and practical strategies are for today.

In this final chapter, I will turn first to the theoretical insights to sketch an 'ideal type' framework as a first step toward understanding the temporal dimension of prolonged international negotiations. As the previous chapters show, existing theoretical frameworks do not sufficiently capture the fluid nature of state preferences and the factors behind the changes in the behaviour of the US and the EU. This is because existing frameworks have traditionally been developed to analyse individual negotiation outcomes, and as a result, too often they have been static in their analysis of prolonged negotiations with multiple agreements.

To address this challenge, this chapter begins in section two by combining our existing theoretical knowledge with the findings of the empirical analysis to identify 'internal' and 'external' factors that explain why the negotiating positions and the type of agreement both parties were prepared to sign changed over time. Section three uses these variables as a basis for considering prolonged international negotiations. The aim is not to develop a formal model in a game-theoretic sense, nor to provide a model with testable if–then propositions. Rather the aim is to sketch out an 'ideal type' framework to improve our understanding of prolonged negotiations. Having done so, section four is able to probe the data further to identify strategies that state and non-state actors can use to influence state behaviour. The final section considers the implications of this study, particularly in so far as it can be generalized to other negotiations. It also highlights the value of considering the temporal dimension of prolonged international negotiations.[1]

HOW AND WHY DO STATE PREFERENCES CHANGE? IDENTIFYING INTERNAL AND EXTERNAL FACTORS

Most studies of international negotiations assume that states have fixed preferences. As discussed in Chapter 2, this is true of the theoretical perspectives in the liberal tradition of international relations. This assumption stems from a focus on individual negotiation outcomes. An analysis of an individual negotiation outcome is forced to assume that preferences are fixed because they are. For example, if one is to ask the preference of the US Government at the negotiations in Kyoto in 1997 it is a fixed preference at a fixed point in time. In assuming that preferences are immutable, much of the literature understandably is concerned with discovering where preferences derive from. The central premise of liberal theory in international relations for instance, is that state preferences derive from domestic and transnational social pressures (Moravcsik, 2008: 236).

However, once the temporal dimension of prolonged international negotiations is taken into account it becomes clear that state preferences are fluid, not fixed. The preferences of states fluctuate over the course of an extended negotiation. The question then, is not simply where do state preferences derive from? But, how and why are they rendered fluid? The evidence from the US and the EU in the climate negotiations is that preferences are rendered fluid by domestic political dynamics, feedbacks from transnational networks, and to some extent, by the behavioural pathways of international regimes.

Reflecting on the evidence from the case studies, I can distinguish two sets of factors that appear important to explaining the observed changes in the behaviour of the US and the EU. Internal factors refer to variables that precipitate a direct shift in state behaviour via the national, international or transnational level. Internal factors include the level of engagement or mobilization of actors, changes in the strategies of actors, or changes in networks of actors, among others. The increased engagement of a treasury department in the domestic discussions is an example of such an internal factor. External factors, on the other hand, refer to variables that shift multiple internal factors and operate independently of the stage of the negotiations, that is, they operate without regard to the movement between the bargaining and the ratification stage. External factors include exogenous shocks, changes in the state of expert knowledge and challenges from other international regimes. For example, a global financial crisis, a nuclear meltdown or a catastrophic hurricane are all possible external shocks that may change state behaviour. In general, the causal chain between an internal factor and a change in state behaviour is shorter and easier to identify than the causal chain between an external factor and a change in state behaviour. Put differently, internal factors are more proximate than external factors, and as a result it is easier to draw inferences from internal factors than it is external ones.[2]

Before discussing each set of factors, several caveats need to be made. First, it is not always possible to draw clear lines around individual factors and there is much overlap, especially given the capacity of one internal factor, such as a new actor, to directly influence another factor, such as the power of a coalition. Second, it should be clear that the intention here is to identify variables that are critical to explaining changes in state behaviour. Hence, I am not interested in factors that, though they may be important to explaining the differences between the behaviour of two or more states, such as the type of domestic political institutions, are not expected to change in the course of prolonged international negotiations. Third, the set of internal and external factors discussed is by no means exhaustive, given it was not possible to consider more than two case studies in depth, in this book. Fourth, and related, the causal chains are suggested chains, they are not causal hypotheses. Further research across a larger population of cases is required before such claims can be made.

Internal Factors

Drawing on the insights of the three theoretical perspectives and the data collected in the preceding empirical chapters, the first and most important internal factor is the level of engagement or mobilization of actors. In prolonged negotiations when actors mobilize there is the potential for new networks between actors and changes in the distribution of the power and preferences of coalitions. As Schattschneider (1960) first pointed out, which stakeholders are mobilized and which are not matters because it affects the balance of forces between actors. As a result, as actors engage and disengage it will create the conditions for new winning and veto coalitions to emerge at the domestic, international and transnational level (Sebenius, 1984: Ch. 6). This might mean the intervention of a treasury department into bureaucratic debates (Allison, 1971), or a new environmental NGO into international discussions (Newell, 2000), or a business group engaging at the transnational level (Keck and Sikkink, 1998). Each new actor could directly precipitate a shift in state behaviour.

This was abundantly clear in both case studies. For example, in the US in the early 1990s economic agencies and business groups had not engaged in the climate discussions and the preferences of environment agencies and environmental NGOs dominated. Yet by the mid-1990s this was changing as business groups, in particular, actively mobilized to oppose an agreement at Kyoto. In the EU too, the dominance of a progressive coalition of environmental interests in the 1990s reflected the fact that other actors were simply not mobilized for most of the Kyoto phase of negotiations, this began to change in the lead up to The Hague in 2000. In short, without allowing for the engagement of actors to vary over time, any understanding of prolonged international negotiations will fail to fully explain state behaviour and negotiation outcomes.

Second, actors make strategic choices about where and how to negotiate. If they choose a new strategy this could shift state behaviour and the outcome. For example, some authors use the term 'forum-shopping' (Braithwaite and Drahos, 2000) or 'different pathways' (Risse-Kappen, 1995b) to describe how actors take actions in different forums or at different levels to influence state behaviour and the outcome of a negotiation. For example, traditionally less powerful actors, such as environmental NGOs, may decide to shift their activities to the international level because of limited access to government at the domestic level, whereas strong actors may not need to do so because of their powerful position at home. Or, as we saw in the US, business groups may supplement their domestic lobbying by engaging in transnational actions as well. This was the strategy of the Climate Council, which was said to represent oil and gas interests in the US. Led by Don Pearlman, a lawyer and former member of the Reagan administration, it formed alliances with OPEC countries, principally Saudi Arabia, who had a similar interest in seeing the negotiations stall.

Third, if we consider the state as a networked entity where the state acts as an agent for the interests of non-state actors and other actors act as agents for states, variations in these networks could influence state behaviour. For example, the importance of 'policy networks' is well established in the governance literature where 'formal institutional and informal linkages between governmental and other actors' shape state behaviour (Rhodes, 2006: 246). They do so by limiting participation in the policy process; deciding which issues will be included and excluded from the policy agenda; shaping the behaviour of actors through the rules of the game; and, among other things, privileging certain interests over others (Rhodes, 2007: 1251). Of course, it is not only national level networks but transnational ones as well that are important (Keohane and Nye, 1974; Risse-Kappen, 1995b; Slaughter, 2004). When these networks change, as they naturally do in the course of prolonged negotiations, so too will state behaviour. This was evident in the case studies. For instance, when new actors mobilized and participated in the policy networks, like the economic agencies in the US, new issues were included on the policy agenda, such as the potential economic costs of the US emissions target. The same was true in the EU when US environmental NGOs began to work closely within the European Commission on an emissions trading policy, a policy which had previously been off the agenda.

Fourth, the mobilization of actors and changes in networks mean the possibility of new coalitions. The two-level intergovernmental framework showed how important the distribution of domestic coalitions is to understanding a state's position and its willingness to sign an agreement (Putnam, 1988). In fact, much of the behaviour of the US and the EU throughout the 1990s can be accounted for by the variations in influence of progressive and regressive coalitions, as discussed in the last chapter. Changes in the distribution of the power and preferences of these coalitions have the potential to directly shift state behaviour.[3] However, in multilateral negotiations it is not just the possibility of new coalitions at the domestic level, but at the international and transnational level as well, that may shape outcomes. As international negotiations drag on, new winning or veto coalitions can emerge to re-shape the dynamics of the negotiations and with it the outcome. For example, at COP 1 in Berlin in 1995, a temporary winning coalition of progressive developing countries and the EU was formed, the so-called 'Green Group', which helped to make the Berlin Mandate possible. Although there was little evidence to indicate that changing transnational coalitions influenced US and EU behaviour, the possibility for such phenomena certainly exist in prolonged negotiations.

While the internal factors identified so far are critical, a key question in the case of prolonged international negotiations is; why do these factors change? In other words, why do new actors mobilize? Why do actors change their strategies? Why do networks among actors change? And, what is it that shifts the distribution of coalitions over time? The short answer is that these changes are the function of the following internal factors: domestic political incentives, the stage of the negotiations and the preferences of the COGs.

Domestic political incentives and the stage of the negotiations are interrelated. As negotiations progress the domestic political incentives for government agencies, environmental NGOs, business groups, or COGs will change and with it their level of engagement. The political incentives for these actors will in turn be a function of how they perceive the tangible costs and benefits of the agreement under negotiation which, as others have pointed out, is directly related to the stage of the negotiations. For example, in elaborating on the two-level game, Moravscik (1993) and Evans (1993) note that as negotiations move from the bargaining to the ratification stage the costs and benefits of an agreement become clearer, and as a result, domestic groups will mobilize in defence of their interests. This in turn will bring new actors into the game. In other words, as some actors push *for* an agreement it engages other actors to push *against* (Spector and Zartman, 2003; Spector, 2003).

The evidence from the US and the EU in the climate negotiations shows that in both cases state and non-state actors mobilized as their political incentives changed in response to the changes in the stage of the negotiations. In the US, as negotiations over ratification wore on in the late 1990s, the engagement of economic agencies and business groups increased as the US emission reduction targets were defined and the economic costs of meeting such targets became tangible. In contrast, in the early 1990s a progressive coalition of environmental interests was able to push their agenda relatively unopposed because negotiations concerned a tentative agreement where the costs and benefits were unclear. It was similar in the EU, where state actors in particular, such as DG Industry in the Commission, acknowledged that it was not until after Kyoto, that is the ratification stage, that officials in these agencies started paying attention.

Finally, as discussed in the last chapter, COGs are crucial to explaining the type of agreement that states are willing to sign (Putnam, 1988; Evans, 1993; Moravcsik, 1993). However, in protracted negotiations there is the potential for changes in the preferences of COGs if there is a change in government, or a change in the beliefs or political incentives of an existing COG. While a change in government is possible in shorter negotiations, it is almost inevitable in protracted negotiations and can lead to a fundamental change in the preferences of the COG. For instance, the electoral success of the Green Party in Europe in the late 1990s and the resulting shift in the composition of the Environment Council directly influenced the preferences of the Council, which acted as the COG for the EU. Yet even when there is no change in government, there is the potential for the preferences of a COG to change with a change in their beliefs or political incentives. For example, when voters are not attentive, it can be expected that politicians will weigh more heavily the voices of organized interest groups on either side of the issue (Harrison and Sundstrom, 2007). As voter interest changes so will the domestic political incentives of the COG.

In the same fashion, the capacity of COGs to manipulate these domestic constraints will also vary over time. Evans (1993) concludes that as international negotiations move from the bargaining to the ratification stage the relative autonomy of the COG to manipulate these pressures decreases. This is because as discussions focus on ratification and tangible costs and benefits, more actors mobilize to advance or defend their interests and hence the constraints on the COG increase. Using the two-level perspective, Trumbore (1998) makes a similar observation that the higher the intensity of the issue under negotiations the greater the likelihood that public preferences will constrain decision-makers. Such variations are likely to become more acute the longer the negotiations. For example, in the US, the preference of the White House, particularly Vice-President Al Gore, was for an ambitious climate agreement based on his own personal beliefs, which in the early 1990s were in line with his domestic political incentives. However, by the time of the negotiations in 2000, domestic actors had mobilized in opposition to an international agreement, and his capacity to manipulate these pressures were limited by a new set of domestic political incentives as a result of his bid for the presidency.

External Factors

Whereas internal factors precipitate a direct shift in state behaviour, external factors, as noted, indirectly shift state behaviour by re-shaping the context in which the negotiations take place. External factors are independent of the stage of the negotiations, that is, they operate without regard to the movement between the bargaining and the ratification stage. However, the longer causal chain between an external factor and a change in state behaviour means that it is not always possible to specify its precise effect.

First, exogenous shocks, or events, have the potential to transform the context in which international negotiations take place. Although no such event occurred in the period of the case studies, others have noted the role that exogenous shocks can play in shifting state behaviour (Zartman, 2003). The most common pathway is where a dramatic event, or series of events captures the imagination of mass publics, after media organizations dramatize the event, and state actors are forced to act to placate the public and the media (Braithwaite and Drahos, 2000; Downs, 1972). The Bhopal accident in 1984 and Chernobyl accident in 1986 are classic cases of exogenous events that catalysed mass publics and forced states to act both domestically and internationally. In the case of climate change, a series of events beginning with an extraordinary heat wave in the US in 1988 along with a congressional inquiry in which NASA scientist James Hansen famously stated that, 'the earth is warmer in 1988 than at any time in the history of instrumental measurements' helped to catalyse the American public and set the stage for the international negotiations, which culminated in the Rio Earth Summit in 1992 (Bodansky, 1993). Such events, which are more likely the longer the negotiation, can shift multiple internal factors as new actors mobilize in response to new political incentives, which in turn affect the distribution of coalitions and so on. In short, an exogenous shock will indirectly shift state behaviour.

A second external factor is the state of expert knowledge among policy elites. As discussed in Chapter 2, epistemic communities, which have an authoritative claim to policy-relevant knowledge within a specific domain or issue-area, can promote organizational learning by helping to create shared understandings in their specialized field (Haas, 1992). In prolonged international negotiations, knowledge within an epistemic community can filter through to change the state of expert knowledge among policy elites. The evolution in the state of knowledge is an external factor because it develops independently of political incentives and the stage of negotiations, yet it has the potential to shift multiple internal factors simultaneously and, consequently, state behaviour. One pathway by which this can occur is by affecting the personal beliefs of statesmen and in turn, their preferences which, as we have seen, are critical to explaining changes in state behaviour (Goldstein and Keohane, 1993). Another is by permeating public opinion so as to shift domestic political incentives.

While it is difficult to measure such effects, the evidence from both cases suggests that this external factor was in play. As an epistemic community, the IPCC was critical to establishing a consensus among policy elites in the US and the EU that there was a discernible human influence on the climate. This was evident both in interviews with policy elites and public statements of government leaders. For example, in the late 1990s environment officials in Germany, Belgium, and Sweden all claimed that they pushed for positions 'that were as close as possible to what the science was calling for' (EU-32; EU-19; EU-26). Similarly in the US, the Vice-President actively pushed the administration toward an agreement in Kyoto based upon his knowledge of the science and his belief that the 'Clinton administration wanted to be on the right side of history' (US-34). This consensus among policy elites did not exist in the 1980s.

Finally, other international regimes can act as an external factor. Zartman (2003: 29–31) has pointed out that competing efforts to deal with aspects of the same problem in overlapping geographic or functional areas occur at the intersection of various regimes. As a result, exogenous challenges can come from other regimes via the same behavioural pathways discussed in Chapter 2. As we have seen, there is evidence to suggest that the international ozone regime provided learning opportunities for EPA officials, which in time were grafted onto domestic and international climate discussions. Further, although there was no evidence during the 1990s that the international trade regime shifted state behaviour in the climate negotiations, there is a growing body of literature to suggest that this will likely occur, if it has not already. (Chambers, 2001; Charnovitz, 2003; New Economics Foundation, 2003; Stokke, 2004; Zhang, 2009). More recently, a collection of studies has highlighted the 'strategic linkages' between the climate regime and other regimes, such as those to combat desertification and protect biological diversity, among others (Jinnah, 2011).

TOWARD AN 'IDEAL TYPE' FRAMEWORK: AN IMMATURE AND MATURE GAME

Having identified a set of internal and external factors that have the potential to shift state behaviour, I am now in a better position to sketch out an 'ideal type' framework for analysing prolonged international negotiations. As seen above, the evidence does not support any simple conclusion that state behaviour shifts in response to one particular factor. Rather, it suggests that the type of agreement a state is willing to sign will change based on the complex interaction of multiple internal and external factors. But this begs the questions: how and when?

To provide a clearer picture, this section suggests that state behaviour can be located along a continuum, of which the opposite ends are an immature game and mature game. Of course, state behaviour will rarely, if ever, conform with either of these extreme positions. Yet it is hard to think more critically about the temporal dimension of prolonged international negotiations without making these types of abstractions. Indeed, they might equally be thought of as 'ideal types', that is, a construct derived from observable reality although not conforming to all of the characteristics because of deliberate simplification (Burger, 1976). In other words, the very concept of an ideal type acknowledges that the real world will only approximate, rather than fully mirror an ideal type of state behaviour.

Before elaborating on what immature and mature games involve, it is important to clarify several points. First, the 'ideal types' are used to characterize the position of a state in the negotiations, not to characterize the international negotiations as a whole. In other words, it is not the international climate change negotiations that are mature or immature but the position of a party to the negotiations. For example, the position of the US or the EU might resemble certain elements of an immature or mature game, but not the negotiations as a whole. Because each ideal type is linked to the stage of the negotiations it is likely that the movement of a state along the continuum will parallel that of other states. That being said, states will move at different speeds based on their domestic, international and transnational interactions, and it is entirely possible than some states will remain in an immature game while others have matured. Second, in reality the types of state behavior are non-binary because in the real world there is unlikely to be a complete distinction between the characterizations of state behavior as an immature game or a mature game. Third, a state's movement along the spectrum between both positions is non-linear. State behavior does not necessarily move in one direction from an immature game to a mature game or vice versa. Instead, state behavior can oscillate between ideal types over time.

An Immature Game

A state will be involved in an immature game when:

(i) The international negotiations are in the 'bargaining phase';
(ii) The costs and benefits of the tentative agreement are intangible;
(iii) Interested actors are minimally mobilized;
(iv) The COG has considerable capacity to manipulate domestic constraints; and
(v) No external factors are in play.

An immature game refers to a position in which many or most of the interested state and non-state actors, with which a state interacts via the domestic, international and transnational level are not mobilized. This is because there are limited political incentives to get involved in the bargaining phase, given negotiations are about a tentative agreement where the costs and benefits of a particular outcome are not yet tangible. As a result, COGs will have a greater capacity to manipulate domestic constraints according to their preferences. Take the example of government actors in the inter-agency discussions. In an immature game very few of these actors are engaged. The senior ranks of government agencies have limited knowledge or interest in the issue because at this stage they will not be directly affected by the negotiated outcome, and competing policy priorities mean that their attention and resources are devoted elsewhere. The same is true of non-state actors such as environmental NGOs and business groups.

This game will be further characterized by the absence of external factors. In an immature game, no exogenous shocks will have taken place to catalyse the public and the media. The state of expert knowledge among policy elites will be immature. Epistemic communities will have only begun to form, or where they have formed, they will not have established a consensus among policy elites about the problem that was different to the state of knowledge before the negotiations began. Other international regimes that are engaged in competing efforts to deal with aspects of the same problem will not have provided any exogenous challenges to actors engaged in the protracted international negotiations.

Finally, the elements of an immature game will impact the likelihood of an agreement and the type of agreement that is reached. States negotiating in immature games will have smaller win-sets. To be sure, because very few interested actors are mobilized a state will have homogeneous interests. While Putnam argues that this will make agreement less likely because the smaller the win-set the less likely it is to overlap with another party's win-set, this will not always hold (Putnam, 1988: 437–8). Putnam claims that a state that is firmly committed to a single policy will be less able to strike a deal internationally than one that is internally divided (Putnam, 1988: 444–5). For example, if the policy of Country A is to support energy companies because they are the only group mobilized, and the policy of Country B is to support emission reductions because environmental NGOs are the only group mobilized it is unlikely that countries A and B will be able to reach an agreement. If however, country B is internally divided because environmental interests and energy companies are mobilized it is possible that the win-sets of both parties will overlap via the preferences of energy companies making agreement more likely. However, the likelihood of an agreement is not simply a function of the probability that win-sets will overlap. Agreement is also contingent on the content of the agreement. In other words, Country A and Country B may both have small win-sets, but if environmental NGOs are the only groups mobilized in both countries their win-sets will still overlap. Furthermore, because in an immature game international negotiations are in a bargaining phase and therefore focussed on a tentative agreement, if an agreement is reached, a non-binding agreement that emphasizes procedures and principles will be the norm.

A Mature Game

A state will be involved in a mature game when:

(i) The international negotiations are in the 'ratification phase';
(ii) The costs and benefits of the agreement are tangible;
(iii) Interested actors are completely mobilized;
(iv) The COG has limited capacity to manipulate domestic constraints; and
(v) All external factors are in play.

In direct contrast, a mature game refers to a position in which many or most of the interested actors with which a state interacts via the domestic, international and transnational level are mobilized. This is because there are political incentives to get involved given that the international negotiations have entered the ratification phase, and the costs and benefits of a particular outcome are now tangible. As a result, COGs will have less capacity to manipulate domestic constraints according to their preferences. Again, using the example of government actors in the inter-agency discussions, in a mature game most actors are engaged. The senior ranks of agencies ensure that their agency is aware of the issues being negotiated and devote sufficient attention and resources to formulate and advocate specific negotiating positions because they will be directly affected by the negotiated outcome. The same is true of non-state actors such as environmental NGOs and business groups.

Further, in a mature game, external factors will be felt with full effect. An external event, such as an environmental crisis, or an economic shock, will have occurred to catalyse the public and the media. The state of expert knowledge among policy elites will have matured so there is a consensus about the problem and the need to resolve it. This consensus will have been established by well-developed epistemic communities, which have had close interaction with policy elites and have likely influenced COG preferences as well. Other international regimes will be engaged in competing efforts to address the same problem, which will have a direct effect on the actors engaged in the international negotiations.

Finally, as with an immature game, the elements of a mature game will impact the likelihood of an agreement and the type of agreement that is reached. States negotiating in mature games are likely to have larger win-sets because most state and non-state actors are mobilized. Admittedly, this will not always be the case. When new actors mobilize in a mature game this could also take options off the table. For example, if energy companies mobilize to form a veto coalition they could reduce the size of the win-set. That being said, when win-sets are larger, again, according to Putnam, agreement will be more likely because the larger the win-set the more likely it is to overlap with another party's win-set (Putnam, 1988: 437–8). However, as discussed, the likelihood of agreement is also contingent on the content of the agreement, so larger win-sets do not always increase the prospect of agreement.

To this set of complex relationships one more needs to be added. The paradox of a larger win-set is that negotiators will be less powerful than those with a small win-set. If a negotiator has a large win-set he can be 'pushed around' by other negotiators, yet if his win-set is small he will be at a bargaining advantage because he can respond that: 'I'd like to accept your proposal, but I could never get it accepted at home' (Putnam, 1988: 440). Furthermore, because international negotiations are in a ratification phase where negotiations are about a legal instrument, if an agreement is reached, a legally binding agreement that includes some implementation requirements will be the norm.

Applying the Ideal Types

In the real world neither of the polar positions of an immature or mature game will be commonly realised. One is not likely to discover many cases replicating a mature game where almost all state and non-state actors are mobilized because each has a political interest to get involved given that the possible agreement will impinge directly on their interests, and all the external factors are in play. For example, there may be cases where all interested actors are mobilized and yet no exogenous shock has taken place. In other words, it will be rare, though not impossible, for a state to perfectly match either ideal type. Nevertheless, the principal aim here is to highlight certain elements about the phenomenon of prolonged international negotiations in order to help improve our understanding of such negotiations. Accordingly, I now turn to consider the ideal types of an immature and mature game in light of the cases of the US and the EU in the international climate negotiations.

In the early 1990s negotiations, both the US and the EU shared many of the characteristics of an immature game. Most state and non-state actors were not mobilized. While government actors, such as environment and energy agencies were actively engaged, along with environmental NGOs, other key actors were not. In the US, the traditionally powerful economic agencies were not mobilized. It was the same in the EU, economic and industry ministries in the member states were not involved in internal discussions and for almost all directorate-generals in the Commission, climate change was not high on their agenda. The same was true for business groups, other than the exceptions of the BTU tax debate in the US and the carbon tax debate in the EU. As a result, the COGs, though there is more evidence for this in the US than the EU, had sufficient autonomy to manipulate domestic constraints according to their preferences.

This reflected two things. First, there were few domestic political incentives for actors to engage at this stage of the negotiations. The negotiations in Berlin promised only to provide the road map for negotiating a legally binding instrument in the future. In other words, negotiations were still in the bargaining phase. This meant that the interests of most actors would not be directly affected by the outcome because the costs and benefits of any agreement remained intangible. Second, external factors had not played a role in the early 1990s. Since the Kyoto phase of negotiations had begun there were no exogenous shocks and there were no external challenges from other regimes, other than some limited influence from the international ozone regime. The only external factor that was in play was the state of expert knowledge among policy elites about the problem of climate change. Yet by 1995, the consensus among US policy elites in particular was not firmly established and doubts remained about the need to act.

In contrast, by the mid-to-late 1990s, the US, and to a lesser extent the EU, conformed to some of the characteristics of a mature game. For example, in the US, the senior ranks of the economic agencies, which had only started to engage prior to Kyoto, along with key business groups, were actively engaged in the discussions by the time of The Hague. In the EU too, there was a small shift among the economic, energy and industry agencies, though not yet at the senior levels. In addition, European business groups became more involved as 'they saw EU policy crystallizing around certain policies that may hurt their interests', such as emissions trading. Again, these changes were the result of a shift in political incentives as the negotiations moved from the bargaining to the ratification stage. In short, as the implications of the Kyoto Protocol became apparent more actors mobilized. This in turn restricted the capacity of the COGs to manipulate the domestic constraints. This was especially evident in the US where Vice-President Al Gore was forced to temper his climate change ambitions.

However, even by 2000, the US and the EU could not be said to represent a perfectly mature game. Some actors were not fully mobilized, for example, EU business groups were not very active. In addition, neither the US nor the EU had experienced an exogenous shock to catalyse the public, and the climate regime had not been challenged by another international regime. That said, there had been a marked shift in the state of expert knowledge among policy elites. The reaction of US policymakers to the release of the IPCC Second Assessment Report in December 1995 was the most prominent indicator of the growing influence of the epistemic community on policy elites. Similarly, interviews with EU officials, such as members of the 'true believers' in the Commission, and those that followed after Kyoto, indicated the extent to which scientific arguments had penetrated elite policymaking and with it the EU negotiating position. In sum, while both the US and the EU positions shared more characteristics of a mature game, by 2000, as would be expected neither case completely conformed to an ideal type.

While an ideal type framework can help us to understand how and when internal and external factors affect state behaviour, it is still not clear why this matters. Put differently, how can this framework help actors to influence state behaviour? In the next section I attempt to answer this question by drawing on the constructs of an immature and mature game to probe the data further to identify strategies that state and non-state actors can use to shape international outcomes in prolonged international negotiations.

STRATEGIES TO INFLUENCE STATE BEHAVIOUR IN PROLONGED INTERNATIONAL NEGOTIATIONS

The concept of an immature and mature game highlights the complex processes that shape prolonged international negotiation outcomes. It is no surprise then, as discussed in Chapter 2, that international relations theorists, especially in the liberal tradition, argue that international outcomes are shaped by more than the economic and security interests of the state, as realists first argued. And, that the literature increasingly employs terms such as 'governance' and 'networks' to capture the complex interdependencies between different types of actors.

In such a world, there are strategic opportunities for highly networked actors to influence state behaviour by making strategic choices at the domestic, international or transnational level to mobilize other actors, establish coalitions, manipulate COG preferences, and in turn, shape international negotiation outcomes. In other words, to engage in what I refer to as 'constructive management'. This recognizes that international outcomes are determined not only by the structural constraints of the international system, as realists envisage, but that state preferences matter to international outcomes, as liberals argue, and that these preferences can be socially constructed, as constructivists would contend.

Further, because preferences are fluid, in a long negotiation actors have considerable agency to influence state behaviour. This is especially so in an immature game given the game itself is fluid, networks are open and coalitions are not highly structured. Even in a mature game, which is more structured and likely to be dominated by powerful actors, weak actors are not without agency. The fact that preferences are fluid in a long negotiation means that actors will always have a degree of agency no matter how weak they are. Some actors may bleakly view a mature game as a triumph of power, and in many cases it will be, but in a long negotiation actors will always have opportunities to shift the preferences of states.

In this section I identify strategies that traditionally weak actors, such as environmental NGOs or environment departments, can employ to steer international negotiations by constructively influencing the negotiating positions of states and the type of agreements they are willing to sign. These strategies could, of course, be employed in shorter one-off negotiations, but they are likely to be uniquely effective when they are used to exploit the strategic opportunities that arise in long negotiations. Accordingly, the choice of strategy will depend not only on the goals of the actor, but importantly in the context of prolonged international negotiations, it will depend on which elements of an immature or mature game are present in the negotiations. In other words, some strategies will be most effective when key elements of an immature game are present and others when a state's position more closely represents a mature game. Each strategy represents an attempt at constructive management.

Four Strategies for Exploiting the Elements of an Immature Game

Exploiting the minimal mobilization of interested actors

In an immature game there is a unique strategic opportunity for traditionally weak actors to dominate discussions because more powerful actors are not mobilized. As discussed above, which actors are mobilized and which are not matters because it affects the balance of forces between actors. This is particularly pertinent in international environmental negotiations where key actors – environment agencies and environmental NGOs – are often the weakest actors. For example, in the US and the EU, the dominance of a progressive coalition of environmental interests in the early 1990s can be largely attributed to limited mobilization of other powerful actors, such as treasury departments or fossil fuel companies.

Accordingly, when interested actors are not mobilized weak actors should actively engage in the discussions as early as possible to exploit the circumstances. In other words, weak actors should go in hard and early to influence state behaviour. For state actors this means that the senior ranks of an agency should engage with the issues being negotiated and devote sufficient attention and resources to formulate and advocate specific negotiating positions, which are likely to have more influence in the absence of alternatives presented by other actors. The same will be true for non-state actors whose positions both on questions of policy and politics will be largely uncontested.

Weak actors who exploit these circumstances will therefore have a unique opportunity to frame the discussions. As discussed in Chapter 2, one of the most effective ways that actors can affect state behaviour is by strategically framing debates to draw attention to their concerns. As other researchers have shown, the actor or coalition of actors that succeeds in establishing a frame that is consistent with its goals is likely to reap the greatest gains from negotiations (Braithwaite and Drahos, 2000; Joachim, 2003; Odell and Sell, 2006). For example, Sell (2003: 182) has argued that 'ideas that are seen as legitimate, appropriate, or correct have a better chance of prevailing' in policy contests.

An interesting question for weak actors in this position is whether or not to engage with more powerful actors. A case in point, as discussed in Chapter 4, was DG Environment in the European Commission in the lead up to the negotiations in Kyoto in 1997. In its attempt to have the Commission propose an ambitious emission reduction target to take to Kyoto, the 'true believers' within DG Environment chose to 'keep climate change off the radar screen of other people in the Commission' to reduce the likely opposition from other directorate-generals (EU-49). For a time, the strategy proved successful as the senior ranks of DG Energy, DG Industry and DG Economics and Finance 'were not very active' (EU-52). However, the issue could only be kept off the radar for so long, and when DG Environment's proposal was finally put at the special chefs meeting, where members of each cabinet sit down to finalize proposals before they go to the Commissioners, the cabinets of other Commissioners were shocked and the proposal met strong resistance.

This case suggests that the most obvious reason to not engage other actors, that is, to keep them in the dark about the issues under negotiation in order to minimize potential opposition, may not always be the best strategy over the course of a prolonged negotiation. This is especially so if the actors that are not yet engaged are likely to act as a powerful veto group further down the track. A more fruitful approach for weak actors might be to exploit the characteristics of an immature game and inform other actors before they have formed a preference on the issues being negotiated. In other words, to frame the policy debate before other actors have an opportunity to do so. The dominant frame is likely to be particularly influential at this point given the limited knowledge other actors have about the issues under negotiation.

Infiltrating and manipulating networks and coalitions

In a networked world, as discussed, where the state acts as an agent for the interests of non-state actors and other actors act as agents for states, highly networked actors have the capacity to shape state behaviour (Rhodes, 2006: 426; 2007). For traditionally weak actors this is often difficult given that policy networks, for example, are often inaccessible. However, in an immature game where a tentative agreement is being negotiated, the costs and benefits of which are not yet tangible and, as a result, interested actors are minimally mobilized, domestic networks and coalitions are likely to be fluid. This will provide a strategic opportunity for weak actors to move in and out of this space with greater ease. As a result, a second and related strategy for weak actors is to infiltrate and manipulate domestic networks and coalitions when they are most fluid.

The success of such a strategy was evident in both the US and the EU prior to the negotiations in Kyoto in 1997. In both cases, environmental NGOs were able to infiltrate and then manipulate orthodox policy networks. In the US, staff from the EDF, who had developed personal relationships with White House advisors, managed to get 'in the room' with key EPA, DoE and White House officials when decisions were made about the US position on emissions trading (US-19). According to interviews with former US officials, EDF, which had played an integral role in developing emissions trading as part of the Clean Air Act

Amendments in 1992, was critical in persuading the Clinton administration to take a similar approach on climate change because it was able to manoeuvre inside key policy networks (US-21; US-34). In the EU, as seen in Chapter 4, environmental NGOs were also successful at infiltrating EU policy networks, as WWF showed with the development of the Triptique approach to burden sharing, which enabled the EU to take a 15 per cent emissions reduction target to the negotiations in Kyoto.

Further, when the structure of the domestic discussions is fluid, state actors also have the capacity to manipulate or steer discussions before more powerful actors mobilize. Prior to 1996 in the US, the CEQ, led by Katie McGinty (Al Gore's confidante), chaired the inter-agency discussions that developed the US negotiating position for COP 1 in Berlin in 1995. However, by 1996, as other interested actors mobilized and the game matured, the NEC successfully lobbied the White House to co-chair the internal discussions with the CEQ. As discussed in Chapter 4, this changed the inter-agency discussions 'considerably' because the EPA, the State Department and the DoE had to contend with the growing input from economic agencies, which were often opposed to their positions (US-25; US-27). In other words, the opportunity to manipulate these networks was restricted once the key elements of an immature game were no longer present.

Targeting the COGs capacity to manipulate domestic constraints

While the preferences of the COG are crucial to explaining state behaviour, their capacity to manipulate domestic constraints is greatest when the international negotiations are in the bargaining phase. The lesson for actors, especially weak actors who have fewer alternative pathways to influence state behaviour, is to target the COG in the bargaining phase. If an actor is successful in influencing a COG's preferences at this point in the negotiations it is more likely that these preferences will be reflected in a state's negotiating position given the greater relative autonomy of the COG.

A COG's preferences will be informed both by personal beliefs and the desire to enhance their domestic political position. Accordingly, actors should target both these avenues to persuade the COG to adopt a position consistent with their interests. First, state and non-state actors should engage directly with the COG and their senior advisors to persuade them of the importance of the issues being negotiated. Of course, this will be easier in an immature game, when such networks will be more porous. One way to do this is to facilitate the flow of expert knowledge about the problem to the COG, as I will discuss in more detail below.

Second, actors should attempt to manipulate the political incentives of the COG in favour of their desired outcome. In international negotiations a good option for weak actors who are unable to mobilize economic resources is to target the reputation of the COG. The evidence, especially from the US case study, suggests several ways that this can be done effectively. One way is to target new governments that appear more susceptible to the costs of reputation because they are trying to stand apart from the previous government. This was certainly the case in the US with the Clinton administration's desire to 'distinguish itself' from the previous Bush administration. Second, COGs are more susceptible to reputational pressures if they have been recently admonished on the international stage, especially over a similar issue such as the provision of a global public good (Downs and Jones, 2002). In the US this was manifest in the months before Kyoto after the US had been ridiculed by the international community for announcing it would not sign an international treaty to ban land mines. Third, and unsurprisingly, a COG's domestic reputation is best targeted when they are seeking re-election. In such circumstances the costs of domestic reputation will likely trump concerns about its international reputation. For example, as noted in Chapter 5, prior to The Hague in 2000, the Clinton administration was not concerned about being seen as a pariah on the international stage, instead the greater concern was its domestic reputation in the context of Vice President Al Gore's run for the White House.

Facilitating the flow of expert knowledge to policy elites

The basic premise of constructive management is that state preferences matter to international outcomes, that they are fluid and that these preferences can be socially constructed. In an immature game where COGs and policy elites have not developed firm preferences on the issue under negotiation, weak actors can facilitate the flow of expert knowledge to these actors to inform their beliefs and, in turn, their preferences and negotiating position. Again, this will work best in an immature game because networks are more fluid and COGs have a greater capacity to manipulate domestic constraints based on their beliefs.

Epistemic communities can play a key role in this process. As seen above, epistemic communities can help to create shared understandings among policy elites about the nature of a problem (Haas, 1992). Weak actors can facilitate the flows of such information. State actors should develop networks with key policy elites and arrange for inter-agency discussions to include experts from epistemic communities to help drive consensus. Non-state actors should play a complementary role organizing to brief policy elites in government, and where possible the COG, of the latest developments in the epistemic community's knowledge. International regimes, like the climate change regime, can be used to justify this process to policy elites who are often pre-occupied with competing policy priorities. For example, as discussed in Chapter 4, after the 'Geneva Declaration' in 1996, 'the ticket to entry' for policy elites to inter-agency meetings 'was that everyone had to have a briefing on the science' (US-7).[4]

Four Strategies for Exploiting the Elements of a Mature Game

The key elements of a mature game make it much more difficult for weak actors to influence state behaviour. With international negotiations in the ratification phase, powerful actors are engaged as the tangible costs and benefits of a negotiated agreement are thrashed out. The result for weak actors is that their positions will be increasingly contested and domestic networks and coalitions are likely to be much more structured and closed off to the infiltration and manipulation that may have been possible earlier in the negotiations. Nevertheless, because preferences are fluid in a protracted international negotiation actors will always have a degree of agency to marshal state behaviour. Accordingly, there are strategies which actors can employ to affect a state's negotiating position and the type of agreement it is willing to sign.

Meeting the demand of policy elites for expert knowledge

First, and very briefly, weak actors should persist in facilitating the flow of expert knowledge to policy elites. While this will be more difficult in a mature game, it still has the potential to shift the preferences of the COG and policy elites. One element of a mature game that will advantage this strategy is that demand for information from policy elites is likely to be greater in the ratification stage when discussions have turned to distributing tangible costs and benefits among domestic interest groups. As Haas (1992) argued, the language of science will penetrate politics especially in issue areas with high complexity and uncertainty where there is demand for such knowledge from policymakers.

Exploiting exogenous shocks

Although no exogenous shocks took place in the period of the case studies, they have a very real potential to shift state behaviour by catalysing mass publics and forcing states to act both domestically and internationally. In such an atmosphere actors who have a pre-prepared model to address the crisis will have an enormous appeal to state actors looking for a solution. For example, if a nuclear meltdown forces a state to act, actors with a pre-prepared model to re-regulate energy production and phase out nuclear power will be in a powerful position. Indeed, the key for weak actors is that their influence depends on the power of the model, not on the power of the advocate. As Braithwaite and Drahos (2000: 589) point out, the politics of modelling is not about problems looking for solutions, but solutions waiting for the right problem and moment to justify their implementation. Accordingly, in a mature game where weak actors have been outmanoeuvred by more powerful actors and are closed off from policy networks, an exogenous shock can provide a strategic opportunity, if they have a model in hand, to influence state behaviour and steer the negotiations toward an outcome consistent with their interests.

Leveraging other international regimes

Further in the context of a mature game dominated by powerful interests, weak actors, where possible, should engage with other international regimes to influence state behaviour. As seen above, one international regime, such as the ozone regime, can provide an exogenous challenge to a second regime, such as the climate regime if they are involved in competing efforts to deal with aspects of the same problem (Zartman, 2003). Where this occurs, strategic opportunities may exist for weak actors whose influence has been muted in one international regime to shift their attention to a second regime and use it as leverage. If possible, actors should engage a stronger regime as this is likely to have a greater capacity to provide an exogenous challenge. For instance, environmental NGOs participating in the climate negotiations may seek to affect the rules and boundaries of the international trade regime with the hope of spurring changes in the climate regime. One avenue could be to introduce rules on border tax adjustments, that is, fees levied by carbon-taxing countries on goods manufactured in non-carbon-taxing countries (New Economics Foundation, 2003; Stiglitz, 2006; Stokke, 2004).

Building transnational coalitions

Finally, as the transnational perspective points out 'transnational relations matter in world politics', and as confirmed by the case studies, state behaviour in international relations cannot be understood without taking account of the cross-boundary activities of sub-units of government and non-state actors (Risse-Kappen, 1995b: 280). In a mature game that is structured by powerful interests who tend to dominate domestic networks and coalitions, a good option for weak actors is to develop transnational networks. The evidence from the case studies indicates that transnational networks are most effective at influencing domestic and international policy outcomes when they include state and non-state actors. For example, as seen in Chapters 4 and 5, by the mid-to-late 1990s a transnational network of American and European actors had developed, which comprised economists within the European Commission and EPA officials in the pursuit of emissions trading. They had also teamed up with non-state actors such as the US NGO the Center for Clean Air Policy to advocate emissions trading in Europe. Further, although there is scant evidence of this from the case studies, in a mature game there is the opportunity for weak actors in one country to enrol more powerful actors in another. For instance, European business groups could have networked with economic officials in the US to ensure that they vetoed any international agreement that would prove costly to American business as well as to European business groups.

IMPLICATIONS FOR FUTURE NEGOTIATIONS AND FUTURE RESEARCH

In 2015, the nations of the world will descend on Paris to reach an agreement on what should replace the Kyoto Protocol. The negotiations will be the latest round in the post-Kyoto phase, which since 2006 have failed to agree on a new legally binding international agreement. Following the hype and despair of Copenhagen in 2009, little progress has been made, and if there is no agreement in the French capital the negotiations are likely to stumble on for some time yet.

This is the reality that state and non-state actors in the US and the EU, among many other nations, are grappling with as they try to conclude this phase of prolonged international climate negotiations, which have now stretched for almost a decade. However, time is running out. Indeed it may already be too late to keep average global temperatures to no more than 2°C above pre-industrial levels, the point at which the worst risks of climate change are to be avoided. In 2013, carbon dioxide levels in the atmosphere reached 400 parts per million (ppm) for the first time in human history and if they reach 450 ppm, there is little chance, if any, of avoiding the 2°C guardrail.

In this context, what are the implications of this book for actors in future climate change negotiations? The short answer is the strategies to influence state behaviour and in turn the negotiation outcomes. If state preferences are fluid not fixed in prolonged international negotiations then there are strategic opportunities for highly networked actors to engage in constructive management by employing the strategies outlined in this chapter. Accordingly, weak actors in the US or the EU, such as environmental NGOs or environment departments, can seek, for example, to exploit exogenous shocks, leverage other international regimes and/or build transnational coalitions. Of course, the choice of strategy will depend not only on the goals of the actor, but in prolonged international negotiations, it will depend on which elements of an immature and mature game are present.

In the current environment, a cursory glance at the US and the EU suggest that they share some of the characteristics of a mature game. For example, most of the interested actors are mobilized, such as government agencies, NGOs and business groups, the COGs appear to have limited capacity to manipulate domestic constraints, as is evident with President Obama's inability to pass climate legislation through the Congress, and some external factors remain in play, notably the state of expert knowledge among policy elites in both the US and the EU about the problem of climate change.

What strategies then could be employed to influence the behaviour of the US and the EU and in turn shape international negotiation outcomes? By way of illustration I will consider three of the strategies discussed above. First, actors could exploit exogenous shocks. The US, for example, has already experienced catastrophic hurricanes, like Katrina in 2005, which destroyed New Orleans and hurricane Sandy in 2012, which flooded parts of New York shutting down the subway and cutting power to parts of the city. Such shocks will only become more frequent and intense as the globe warms. The key for actors is to develop models now that states can implement in response to future shocks. This could include plans to reconfigure a city's transport system, re-design US energy systems, or perhaps most importantly, global energy systems. The International Energy Agency has already called for an 'energy revolution' stating that 'it is not an exaggeration to claim that the future of human prosperity depends on how successfully we tackle' the energy sector (IEA 2008).

Second, to the extent that the elements of a mature game are present in the US and the EU, now could be a fruitful time for actors to leverage other international regimes to affect future negotiation outcomes. One avenue could be the G20 – a grouping of 20 major economies, including China, India and Brazil, whose heads of state have met annually since the global financial crisis in 2008. While the G20 is largely focused on economic and financial issues, there is some appetite for it to play a role on climate change. For example, at the G20 meeting in France in 2011 world leaders agreed that new sources of funding needed to be found to address climate change. Given its membership, the G20 has the potential to develop into a powerful regime, with the capacity to provide an exogenous challenge to the climate regime.

Third, in a mature game structured by powerful interests who tend to dominate domestic networks and coalitions, as is likely to be the case in the US and the EU at present, the development of transnational networks could be an effective strategy. For weak actors, especially environmental NGOs, who have failed to influence state behaviour through more orthodox means, such as domestic or international lobbying, transnational networks may prove to be the best alternative. Some NGOs, such as 350.org, which seeks to build a global grassroots movement for action on climate change, are rapidly pursuing this path in the US, Canada and Australia. While it remains to be seen how effective these strategies will be in influencing state behaviour and shaping future negotiation outcomes, the evidence from the case studies indicate that transnational coalitions are likely to be more effective when they include state as well as non-state actors, be they in the US, EU or elsewhere.

If these are the implications for future prolonged international climate negotiations, what can be said beyond that? What are the broader implications of this book, and can they be generalized to other states and other prolonged international negotiations?

In broad terms, this book has shown that states are the most important actors in international negotiations. However, this book does not support a classical realist account of international relations. Yes, the state is the most important actor, but the state is more disaggregated than realists would ever contend, and international negotiations outcomes are shaped by more than the economic and security interests of states. International negotiations now take place in a highly networked age where state and non-state actors have the capacity to influence state behaviour by making strategic choices at the domestic, international or transnational level. This account is closer to the liberal and constructivist schools of thought and consistent with recent pronouncements that terms such as 'global governance' more closely resonate with reality than 'international relations' (Lake, 2010).

However, the fluidity of preferences that characterizes prolonged international negotiations makes it almost impossible to develop the type of parsimonious theory toward which realists, liberals and others in international relations strive. If preferences were like neutrons it would be possible to develop if–then propositions, but in a long negotiation preferences are positively charged one year and negatively charged the next. Accordingly, once it is acknowledged that preferences are not fixed, but fluid in long negotiations, attempts at parsimonious theory may provide elegance, but they will also limit many important insights, which contribute to a greater understanding of prolonged international negotiations.

That said, this book has shown more specifically, that the two-level intergovernmental perspective provides the most convincing explanation of the behaviour of the US and the EU in the international climate negotiations. Although a complete explanation requires the insights of the transnational approach and the international regime approach as well. Second, this book has also shown that existing perspectives do not sufficiently capture the factors behind the changes in US and EU behaviour. Accordingly, this book has identified a general set of internal and external factors distinctive to prolonged international negotiations that explain how and why the negotiating positions and the type of agreement both parties were prepared to sign changed. These have then been used as a basis to sketch an ideal type framework as a first step for understanding the temporal dimension of prolonged international negotiations and prescribing specific strategies that state and non-state actors can use to shape international outcomes.

While these achievements are important, I cannot conclude the analysis of prolonged international negotiations here. First, although the findings of this book can be generalized to theoretical propositions, based on only two case studies there are limits to how far they can be generalized to populations (Yin, 2009: 9). Further research is needed to test these theoretical propositions across a wider range of cases, and across other issue areas. For example, given the protracted nature of the Uruguay Round and the Doha Round, the international trade negotiations would likely provide a rich body of data for future analysis of prolonged international negotiations. Second, as discussed in Chapter 1, the aim of this book was not to develop a formal model in a game-theoretic sense, instead the research stressed the inductive generation of propositions and generalizations about state behaviour in prolonged international negotiations. Accordingly, additional empirical research across a larger population of cases combined with rigorous deductive work is needed for formal models to be developed.

Notwithstanding, this book has shown the value of analyzing variations in state behavior across prolonged international negotiations. In other words, it has demonstrated the insights that are gained from looking at individual negotiation outcomes as one would a still photo, and then putting them together to create a moving picture. To the extent that this moving picture has improved our understanding of the temporal dimension of prolonged international negotiations, it will improve the likelihood that states are able to reach agreement on some of the most critical problems facing the globe, such as climate change, or future negotiations around food, water or security issues.

NOTES

1. This chapter is based on a paper published in *International Negotiation* (Downie, 2012).
2. I am indebted to Lawrence Susskind for suggesting the idea of internal and external factors to think about the variations in prolonged international negotiations. Personal communication November 2010.
3. This is largely consistent with liberal theory, which stresses that the nature of domestic political representation helps to determine whose social preferences dominate state policy. As a result, policy tends to be biased in favour of the governing coalitions or powerful domestic groups favoured by representative institutions, be they armies, bureaucracies or societal groups (Moravcsik, 2008).
4. There are other examples of international regimes legitimising this process. For instance, the Barents Sea fisheries regimes helped to institutionalise the role of scientists in the decision-making process (Young, 1999: 263–64).

References

Abbott, K. 2011. 'The transnational regime complex for climate change'. *SSRN Working Paper.* Tempe: Arizona State University.

Adler, E. 2002. 'Constructivism and international relations'. *In:* Carlsnaes, W., Risse, T. and Simmons, B. (eds) *Handbook of International Relations*, London: Sage Publications.

Agrawala, S. and Andresen, S. 1999. 'Indispensability and indefensibility? the United States in the climate treaty negotiations'. *Global Governance*, **5**, 457–82.

Albright, M. 1998. 'Earth day 1998: global problems and global solutions', *remarks at Earth Day 1998 at the National Museum of Natural History.* Washington DC.

Aldy, J. 2004. 'Saving the planet cost-effectively: the role of economic analysis in climate change mitigation policy'. *In:* Lutter, R. and Shogren, J. (eds) *Painting the White House Green: Rationalizing Environmental Policy Inside the Executive Office of the President*, Washington DC: RFF Press.

Allison, G. 1971. *Essence of Decision: Explaining the Cuban Missile Crisis*, New York: Harper Collins.

Allison, I., Bindoff, N.L., Bindschadler, R.A., Cox, P.M., De Noblet, N., England, M.H., Francis, J.E., Gruber, N., Haywood, A.M., Karoly, D.J., Kaser, G., Le Quere, C., Lenton, T.M., Mann, M.E., McNeil, B. I., Pitman, A.J., Rahmstorf, S., Rignot, E., Schellnhuber, H.J., Schneider, S.H., Sherwood, S.C., Somerville, R.C.J., Steffen, K., Steig, E.J., Visbeck, M. and Weaver, A.J. 2011. *The Copenhagen Diagnosis: Updating the World on the Latest Climate Science*, Oxford: Elsevier.

Andonova, L., Betsill, M.M. and Bulkeley, H. 2007. 'Transnational climate change governance'. *Amsterdam Conference on the Human Dimensions of Global Environmental Change.* Amsterdam.

Atkinson, R. 1995. 'Climate debate ends on cool note greenhouse gas cutbacks stalled 2 years, *The Washington Post*, 8 April.

Barnett, M. and Sikkink, K. 2008. 'From international relations to global society'. *In:* Reus-Smit, C. and Snidal, D. (eds), *The Oxford Handbook of International Relations*, Oxford: Oxford University Press.

Benedick, R. 1991. *Ozone Diplomacy: New Directions in Safeguarding the Planet*, Cambridge: Harvard University Press.

Bennett, A. 2007. 'Case study methods: design, use, and comparative advantages'. *In:* Sprinz, D. and Wolinsky-Nahmias, Y. (eds), *Models, Numbers and Cases: Methods for Studying International Relations*, Ann Arbor: The University of Michigan Press.

Berry, J.M. 2002. 'Validity and reliability issues in elite interviewing'. *PS: Political Science and Politics*, **35**, 679–82.

Betsill, M. 2008. 'Environmental NGOs and the Kyoto Protocol negotiations: 1995 to 1997'. *In:* Betsill, M. and Corell, E. (eds), *NGO Diplomacy: The Influence of Nongovernmental Organizations in International Environmental Negotiations*, Cambridge: The MIT Press.

Betsill, M. and Bulkeley, H. 2004. 'Transnational networks and global environmental governance: the cities for climate protection program', *International Studies Quarterly*, **48**, 471–93.

Beuermann, C. and Jager, J. 1996. 'Climate change politics in Germany: how long will any double dividend last?' *In:* O'Riordan, T. and Jager, J. (eds) *Politics of Climate Change: A European Perspective*, London: Routledge.

Bodansky, D. 1993. 'The United Nations framework convention on climate change: a commentary'. *Yale Journal of International Law*, **18**, 451–558.

Bodansky, D. 2010. *The Copenhagen Climate Change Conference: A Post-Mortem*, Athens: University of Georgia Law School.

Boulton, L. and Hutton, B. 1997. 'Kyoto climate change talks agree treaty at last minute'. *The Financial Times*, 11 December, p. 1.

Braithwaite, J. and Drahos, P. 2000. *Global Business Regulation*, Cambridge: Cambridge University Press.

Breitmeier, H., Young, O. and Zurn, M. (eds) 2006. *Analyzing International Environmental Regimes: From Case Study to Database*, Cambridge: The MIT Press.

Brown, P. 1995. 'Gummer attempts to break climate conference impasse'. *The Guardian*, 5 April.

Brown, P. 1997. 'Prescott takes lead on climate change', *The Guardian*, 30 October.

Brown, P. 2000. 'US berated for wriggling out of treaty pledges', *The Guardian*, 22 November.

Browne, J. 1997. 'Breaking Ranks', *Stanford Business*, (September), 16–19.

Bryner, G. 2000. 'Congress and the politics of climate change'. *In:* Harris, P. (ed.) *Climate Change and American Foreign Policy*, New York: St Martin's Press.

Bull, H. 1995. *The Anarchical Society: A Study of Order in World Politics*, New York: Columbia University Press.

Burger, T. 1976. *Max Weber's Theory of Concept Formation: History, Laws and Ideal Types*, Durham: Duke University Press.

Busby, J. 2010. *Moral Movements and Foreign Policy*, Cambridge: Cambridge University Press.
Carmody, K. 1995. 'It's a jungle out there: environmental journalism in an age of backlash', *Columbia Journalism Review*, **34**, 40–46.
Carpenter, C. 2001. 'Businesses, green groups and the media: the role of non-governmental organizations in the climate change debate', *International Affairs*, **77**, 313–28.
Cass, L. 2006. *The Failures of American and European Climate Policy*, Albany: State University of New York Press.
Cass, L. 2007. 'The indispensable awkward partner: The United Kingdom in European climate policy'. *In:* Harris, P. (ed.) *Europe and Global Climate Change: Politics, Foreign Policy and Regional Cooperation*, Cheltenham: Edward Elgar.
Chambers, B. 2001. 'International trade law and the Kyoto Protocol: potential incompatibilities'. *In:* Chambers, B. (ed.) *Inter-linkages: The Kyoto Protocol and the International Trade and Investment Regimes*, Tokyo: University Press.
Charnovitz, S. 2003. 'Trade and climate: potential conflicts and synergies'. *Beyond Kyoto: Advancing the International Effort Against Climate Change*, Pew Centre on Global Climate Change.
Checkel, J. 2001. 'Why comply? Social learning and European identity change', *International Organization*, **55**, 553–88.
Christiansen, A. and Wettestad, J. 2003. 'The EU as a frontrunner on greenhouse gas emissions trading: how did it happen and will the EU succeed?' *Climate Policy*, **3**, 3–18.
Clark, B. 1997. 'Gases talks underline EU-US gap'. *The Financial Times*, 5 December, p. 4.
Clemencon, R. 2008. 'The Bali road map: a first step on the difficult journey to a post-Kyoto Protocol agreement', *The Journal of Environment and Development*, **17**, 70–94.
Collier, U. 1997. "Windfall" emission reductions in the UK'. *In:* Collier, U. and Lofstedt, R. (eds) *Cases in Climate Change Policy: Political Reality in the European Union*, London: Earthscan.
Collier, U. and Lofstedt, R. 1997. 'Comparative analysis and conclusions'. *In:* Collier, U. and Lofstedt, R. (eds) *Cases in Climate Change Policy: Political Reality in the European Union*, London: Earthscan.
Committee on Environment and Public Works United States Senate 1997. Reviewing the effects of greenhouse gases on global weather conditions and assessing international policy options to reduce the negative impacts of climate change, 105th Congress: 1st Session.
Cooper, R. 2007. 'Alternatives to Kyoto: the case for a carbon tax'. *In:* Aldy, J. and Stavins, R. (eds) *Architectures for Agreement: Addressing*

Global Climate Change in the Post-Koyto World, Cambridge: Cambridge University Press.

Corell, E. and Betsill, M. 2008. 'Analytical framework: assessing the influence of NGO diplomats'. *In:* Corell, E. and Betsill, M. (eds) *NGO Diplomacy: The Influence of Nongovernmental Organizations in International Environmental Negotiations*, Cambridge: The MIT Press.

Cortell, A. P. and Davis, J. W. 1996. 'How do international institutions matter? the domestic impact of international rules and norms', *International Studies Quarterly*, **40**, 451–78.

Council of Economic Advisors 1998. 'The Kyoto Protocol and President's policies to address climate change: administration economic analysis', Washington DC: White House.

Cushman, J. 1998. 'Study says emissions cut will raise fuel costs', *New York Times*, 9 October.

Cushman, J. and Sanger, D. 1997. 'No simple fight: the forces that shaped the Clinton plan', *New York Times*, 1 December.

Darkin, B. 2006. 'Pledges, politics and performance: an assessment of UK climate policy', *Climate Policy*, **6**, 257–74.

Daugberg, C. and Swinbank, A. 2009. *Ideas, Institutions, and Trade: The WTO and the Curious Role of EU Farm Policy in Trade Liberalization*, Oxford: Oxford University Press.

Delaney, K. 2007. 'Methodological dilemmas and opportunities in interviewing organizational elites'. *Sociology Compass*, **1**, 208–21.

Depledge, J. 2005. *The Organization of Global Negotiations: Constructing the Climate Change Regime*, Sterling, VA: Earthscan.

Directorate-General for Economic and Financial Affairs, 1992a. 'The climate challenge: economic aspects of the community's strategy for limiting CO_2 emissions'. *European Economy*, Brussels: European Commission.

Directorate-General for Economic and Financial Affairs, 1992b. 'The economics of limiting CO2 emissions'. *European Economy*, Brussels: European Commission.

Di Valdalbero, D.R. 2010. *The Power of Science: Economic research and European decision-making: The case of energy and environment policies*, Brussels: P.I.E. Peter Lang.

Doran, P. 2000. 'Upholding the 'Island of High Modernity': the changing climate of American foreign policy'. *In:* Harris, P. (ed.), *Climate Change and American Foreign Policy*, New York: St. Martin's Press.

Downie, C. 2012. 'Toward an understanding of prolonged international negotiations'. *International Negotiation*, **17**, 295-320.

Downie, C. 2013a. 'Transnational actors: nongovernmental organizations, civil society and individuals'. *In:* Harris, P. (ed.) *Routledge Handbook of Global Environmental Politics,* London: Routledge.

Downie, C. 2013b. 'Three ways to understand state actors in international negotiations: climate change in the Clinton years (1993-2000)'. *Global Environmental Politics*, **13**, 22-40.
Downie, C. 2013c. 'Shaping international negotiations from within the EU: sub-state actors and climate change'. *Journal of European Integration*, **35**, 705–21.
Downie, D.L., Chasek, P., Baumert, K., Clark, S., Tosteson, J., Bissell, L., Hjerthen, J., Jayaraman, B., Karkus, E., Leahy, J. and Mulder, G. 1998. European Union views on international greenhouse gas emissions trading. Columbia University School of International and Public Affairs. May 1998.
Downs, A. 1972. 'Up and down with ecology – the "issue-attention cycle"'. *Public Interest*, **28**, 38–50.
Downs, G. and Jones, M. 2002. 'Reputation, compliance, and international law'. *Journal of Legal Studies*, **31**, S95–S114.
Druckman, D. and Mahoney, R. 1977. 'Processes and consequences of international negotiations'. *Journal of Social Issues*, **33**, 60–87.
Earth Negotiations Bulletin 1995. 'Summary of the first conference of the parties to the UN framework convention on climate change: 28 March–7 April 1995'. *International Institute for Sustainable Development*, **12**, 1–11.
Earth Negotiations Bulletin 1996. 'Summary of the second conference of the parties to the UN framework convention on climate change: 8–19 July 1996'. *International Institute for Sustainable Development*, **12**, 1–14.
Earth Negotiations Bulletin 1997. 'Report of the third conference of the parties to the UN framework convention on climate change: 1–11 December 1997'. *International Institute for Sustainable Development*, **12**, 1–16.
Earth Negotiations Bulletin 2000. 'Summary of the sixth conference of the parties to the UN framework convention on climate change: 13–25 November 2007'. *International Institute for Sustainable Development*, **12**, 1–19.
Eisenhardt, K. 1989. 'Building theories from case study research'. *The Academy of Management Review*, **14**, 532–50.
Energy Information Administration 1998a. 'Higher energy prices, cuts in fuel use may be needed to comply with the Kyoto Protocol', *Press Release.* Washington DC.
Energy Information Administration 1998b. 'Impacts of the Kyoto Protocol on US energy markets and economic activity', Washington DC: United States Government.
Enia, J. 2009. 'Sequencing negotiating partners: implications for the two-level game?', *Negotiation Journal*, **25,** 357–83.
European Commission 1993. 'The fifth EC environmental action programme', Brussels: European Commission.

European Commission 2000. 'Green paper on greenhouse gas emissions trading within the European Union', *Green Paper*, Brussels: European Commission.

European Environment Council 1997a. 1990th Council meeting environment, *Press Release*, Brussels. 3 March.

European Environment Council 1997b. 2017th Council meeting environment, *Press Release*, Brussels. 20 June.

European Environment Council 1997c. 2062nd Council meeting environment, *Press Release*, Brussels. 16 December.

European Environment Council 2000. 2302nd Council meeting environment, *Press Release*, Brussels. 7 November.

European Report 2000a. Environment: firm stance from ministers on climate change, Brussels: European Report. 8 November.

European Report 2000b. 'Environment: Parliament urges EU to take lead in combating climate change', 28 October.

Evans, P. 1993. 'Building an integrative approach to international and domestic politics: reflections and projections'. *In:* Evans, P., Jacobson, H. and Putnam, R. (eds.) *Double-Edged Diplomacy: International Bargaining and Domestic Politics*, Berkeley: University of California Press.

Fialka, J. 1997. 'Efforts to curb global warming is tied to higher energy prices in two studies', *The Wall Street Journal*, 16 July.

Fidler, S. and Houlder, V. 2000. 'Europeans defend rejection of climate deal', *Financial Times*, 28 November.

Fineman, H. and Breslau, K. 1997. 'Gore Feels the Heat', *New York Times*, 27 October.

Fisher-Vanden, K. 2000. 'International policy instrument prominence in the climate change debate'. *In:* Harris, P. (ed.) *Climate Change and American Foreign Policy*, New York: St Martin's Press.

Foley, J. 1999. 'US Reaction to EU move to limit emission trading', *US Department of State*, May 17.

Frankel, J. 2001. 'You're getting warmer: the most feasible path for addressing global climate change does run through Kyoto', *Fondazione Eni Enrico Mattei*, Milan, Italy.

Garnaut, R. 2008. 'Keynote address to the climate change and social justice conference', *University of Melbourne*, Melbourne.

Gelbspan, R. 1997. *The Heat Is On: The High Stakes Battle over Earth's Threatened Climate*, Reading: Addison-Wesley.

Giorgetti, C. 1999. 'From Rio to Kyoto: a study of the involvement of non-governmental organizations in the negotiations on climate change', *NYU Environmental Law Journal*, **7**, 201–245.

Goldstein, J. and Keohane, R. 1993. 'Ideas and foreign policy: an analytical framework'. *In:* Goldstein, J. and Keohane, R. (eds) *Ideas and Foreign Policy: Beliefs, Institutions and Political Change*, Ithaca: Cornell University Press.

Gough, C. and Shackley, S. 2001. 'The respectable politics of climate change: the epistemic communities and NGOs', *International Affairs*, **77**, 329–45.

Grubb, M., Koch, M., Thomson, K., Munson, A. and Sullivan, F. 1993. *The 'Earth Summit' Agreements: A Guide and Assessment*, London: Earthscan.

Gulbrandsen, L. and Steinar, A. 2004. 'NGO influence in the implementation of the Kyoto Protocol: compliance, flexibility mechanisms, and sinks', *Global Environmental Politics*, **4**, 54–75.

Haas, P. 1989. 'Do regimes matter? epistemic communities and Mediterranean pollution control', *International Organization*, **43**, 377–403.

Haas, P. 1992. 'Introduction: epistemic communities and international policy coordination'. *International Organization*, **46**, 1–35.

Hafner-Burton, E., Kahler, M. and Montgomery, A. 2009. Network analysis for international relations, *International Organization*, **63**, 559–92.

Haigh, N. 1996. 'Climate change policies and politics in the European community'. *In:* O'Riordan, T. and Jager, J. (eds) *Politics of Climate Change: A European Perspective*, London: Routledge.

Harrison, N. 2000. 'From the inside out: domestic influences on global environmental policy'. *In:* Harris, P. (ed.), *Climate Change and American Foreign Policy*, New York: St Martin's Press.

Harrison, K. and Sundstrom, L. 2007. 'The comparative politics of climate change', *Global Environmental Politics*, **7**, 1–18.

Hasenclever, A., Mayer, P. and Rittberger, V. 2000. 'Integrating theories of international regimes', *Review of International Studies*, **26**, 3–33.

Hatch, M. 2007. 'The politics of climate change in Germany: domestic sources of environmental foreign policy'. *In:* Harris, P. (ed.), *Europe and Global Climate Change: Politics, Foreign Policy and Regional Cooperation*, Cheltenham: Edward Elgar.

Helm, M. 1997. 'Gore feels heat to attend talk on global warming', *The Times Picayune*, 23 November.

Herbert, J. 2000. 'Nations discuss proposals on Kyoto greenhouse treaty', *Pittsburgh Post-Gazette*, 11 November.

Hilzenrath, D. 1993. 'Miscalculations, lobby effort doomed BTU tax plan', *The Washington Post*, 11 June.

Hurd, I. 2008. 'Constructivism'. *In:* Reus-Smit, C. and Snidal, D. (eds), *The Oxford Handbook of International Relations*, Oxford: Oxford University Press.

International Energy Agency (IEA). 2008, *World Energy Outlook 2008*, Paris: IEA.

IPCC 1995. *IPCC Second Assessment Report: Climate Change 1995*, Cambridge: Cambridge University Press.

IPCC 2013. *Working Group I Contribution to the IPCC Fifth Assessment Report Climate Change 2013: The Physical Science Basis Summary for Policymakers,* Stockholm: IPCC Secretariat.

Jervis, R. 1988. 'Realism, game theory, and cooperation', *World Politics*, **40**, 317–49.

Jinnah, S. 2011. 'Climate change bandwagoning: the impacts of strategic linkages on regime design, maintenance, and death', *Global Environmental Politics*, **11**, 1–9.

Joachim, J. 2003. 'Framing issues and seizing opportunities: the UN, NGOs, and women's rights', *International Studies Quarterly*, **47**, 247–74.

Jonsson, C. and Tallberg, J. 1998. 'Compliance and post-agreement bargaining', *European Journal of International Relations*, **4**, 371–408.

Jordan, A. and Rayner, T. 2010. 'The evolution of climate policy in the European Union: an historical overview'. *In:* Jordan, A., Huitema, D., Asselt, H. V., Rayner, T. and Berkhout, F. (eds), *Climate Change Policy in the European Union: Confronting the Dilemmas of Mitigation and Adaptation?*, Cambridge: Cambridge University Press.

Kanie, N. 2003. 'Leadership in multilateral negotiation and domestic policy: The Netherlands at the Kyoto Protocol negotiation', *International Negotiation*, **8**, 339–65.

Keck, M. and Sikkink, K. 1998. *Activists Beyond Borders: Advocacy Networks in International Politics*, Ithaca: Cornell University Press.

Keohane, R. 1984. *After Hegemony: Cooperation and Discord in the World Political Economy*, New Jersey: Princeton University Press.

Keohane, R. and Nye, J. (eds) 1972. *Transnational Relations and World Politics*, Cambridge: Harvard University Press.

Keohane, R. and Nye, J. 1974. 'Transgovernmental relations and international organizations', *World Politics*, **27**, 39–62.

King, G., Keohane, R.O. and Verba, S. 1994. *Designing Social Inquiry: Scientific Inference in Qualitative Research*, Princeton, Princeton University Press.

Kinzer, S. 1995. 'The world; Europe's leaders see green', *The New York Times*, 9 April.

Kolk, A. and Levy, D. 2001. 'Winds of change: corporate strategy, climate change and oil multinationals', *European Management Journal*, **19**, 501–9.

Krasner, S. 1972. 'Are bureaucracies important? (or Allison Wonderland)'. *Foreign Policy,* Summer**,** 159–79.

Krasner, S. 1983. 'Structural causes and regime consequences: regimes as intervening variables'. *In:* Krasner, S. (ed.), *International Regimes*, Ithaca: Cornell University Press.

Kroll, S. and Shogren, J. 2008. 'Domestic politics and climate change: international public goods in two-level games'. *Cambridge Review of International Affairs*, **21**, 563–83.

Lacasta, N., Dessai, S., Kracht, E. and Vincent, K. 2007. 'Articulating a consensus: the EU's position on climate change'. *In:* Harris, P. (ed.), *Europe and Global Climate Change: Politics, Foreign Policy and Regional Cooperation*, Cheltenham: Edward Elgar.

Lake, D. 2008. 'The state and international relations'. *In:* Reus-Smit, C. and Snidal, D. (eds), *Oxford Handbook of International Relations*, Oxford: Oxford University Press.

Lake, D. 2010. 'Rightful rules: authority, order, and the foundations of global governance', *ISA Presidential Address.*

Leggett, J. 1999. *The Carbon War: Dispatches from the End of the Oil Century*, London: Allen Lane.

Levy, D. 2005. 'Business and the evolution of the climate regime'. *In:* Levy, D. and Newell, P. (eds), *The Business of Global Environmental Governance*, Cambridge: The MIT Press.

Levy, M., Young, O. and Zurn, M. 1995. 'The study of international regimes', *European Journal of International Affairs*, **1**, 267–330.

Loy, F. 2000. 'Testimony of Frank Loy, Undersecretary for Global Affairs', *Committee on Foreign Relations and the Committee on Energy and Natural Resources*, Washington DC.

March, J. and Olsen, J. 1998. 'The institutional dynamics of international political orders', *International Organization*, **52**, 943–69.

McGinty, K. 1993. 'Press briefing by Director of the White House Office of Environmental Policy', Washington DC: White House.

McKeown, T. 2009. 'The big influence of big allies: transgovernmental relations as a tool of statecraft'. *In:* Milner, H. and Moravcsik, A. (eds), *Power, Interdependence, and Nonstate Actors in World Politics*, Princeton: Princeton University Press.

Meckling, J. 2011. *Carbon Coalitions: Business, Climate Politics, and the Rise of Emissions Trading*, Cambridge; The MIT Press.

Milner, H. 1992. 'International theories of cooperation among nations: strengths and weaknesses', *World Politics*, **44**, 466–96.

Milner, H. and Moravcsik, A. (eds) 2009. *Power, Interdependence, and Nonstate Actors in World Politics*, Princeton: Princeton University Press.

Moravcsik, A. 1993. 'Introduction: Integrating international and domestic theories of international bargaining'. *In:* Evans, P., Jacobson, H. and

Putnam, R. (eds), *Double-Edged Diplomacy: International Bargaining and Domestic Politics*, Berkeley: University of California Press.

Moravcsik, A. 2008. 'The new liberalism'. *In:* Reus-Smit, C. and Snidal, D. (eds), *The Oxford Handbook of International Relations*, Oxford: Oxford University Press.

New Economics Foundation 2003. 'Free riding on the climate: the possibility of legal, economic and trade restrictive measures to tackle inaction on global warming', London: New Economics Foundation.

Newell, P. 2000. *Climate for Change: Non-state Actors and the Global Politics of the Greenhouse*, Cambridge: Cambridge University Press.

Nordhaus, T. 2008. 'Getting real on climate change', *The American Prospect*, **19** (12), November.

O'Leary, H. 1993. 'Press briefing by Secretary of Energy Hazel O'Leary', Washington DC: White House.

Oberthur, S. and Ott, H. 1999. *The Kyoto Protocol: International Climate Policy for the 21st Century*, Berlin: Springer.

Oberthur, S. and Tanzler, D. 2007. 'Climate policy in the EU: international regimes and policy to diffusion'. *In:* Harris, P. (ed.), *Europe and Global Climate Change: Politics, Foreign Policy and Regional Cooperation*, Cheltenham: Edward Elgar.

Odell, J. 1993. 'International threats and internal politics: Brazil, the European Community, and the United States 1985–1987'. *In:* Evans, P., Jacobson, H. and Putnam, R. (eds), *Double-Edged Diplomacy: International Bargaining and Domestic Politics*, Berkeley: University of California Press.

Odell, J. 2006a. 'Introduction'. *In:* Odell, J. (ed.), *Negotiating Trade: Developing Countries in the WTO and NAFTA*, Cambridge: Cambridge University Press.

Odell, J. and Sell, S. 2006. 'Reframing the issue: the WTO coalition on intellectual property and public health, 2001'. *In:* Odell, J. (ed.), *Negotiating Trade: Developing Countries in the WTO and NAFTA*, Cambridge: Cambridge University Press.

OECD 1992a. 'Climate change: designing a practical tax system', Paris: Organization for Economic Co-operation and Development.

OECD 1992b. 'Climate change: designing a tradeable permit system', Paris: Organization for Economic Co-operation and Development.

Ott, H., Sterk, W. and Watanabe, R. 2008. 'The Bali roadmap: new horizons for global climate policy', *Climate Policy*, **8**, 91–5.

Peterson, T. 2004. 'The evolution of state climate change policy in the United States: lessons learned and new directions', *Widener Law Journal*, **14**, 81–120.

Pitzl, J. and Stewart, E. 2000. 'The US position on the Kyoto Protocol: senate ratification or not?', Washington DC: Institute for the Study of Diplomacy.

Pizer, W. 2004. 'A tale of two policies: clear skies and climate change'. *In:* Lutter, R. and Shogren, J. (eds), *Painting the White House Green: Rationalizing Environmental Policy Inside the Executive Office of the President*, Washington DC: Resources for the Future.

Pollack, M. and Shaffer, G. 2001. 'Transatlantic governance in historical and theoretical perspective'. *In:* Pollack, M. and Shaffer, G. (eds), *Transatlantic Governance in the Global Economy*, Rowman and Littlefield Publishers: Lanham.

Porter, G., Brown, J. and Chasek, P. (eds) 2000. *Global Environmental Politics*, Boulder Colorado: Westview Press.

Priest, D. 1997. 'Clinton directive on mines: new form, old function', *The Washington Post*, 24 September.

Pronk, J. 2000. 'Note by the President of COP 6'. UNFCCC: The Hague.

Putnam, R. 1988. 'Diplomacy and domestic politics: the logic of two-level games', *International Organization*, **42**, 427–60.

Raustiala, K. and Bridgeman, N. 2007. 'Nonstate actors in the global climate regime', *Public Law and Legal Theory Paper Research Series*, Research Paper No. 07–29.

Rhodes, R. 2006. 'Policy network analysis'. *In:* Moran, M., Rein, M. and Goodin, R. (eds) *The Oxford Handbook of Public Policy*, Oxford: Oxford University Press.

Rhodes, R. 2007. 'Understanding governance: ten years on', *Organization Studies*, **28**, 1243–64.

Ringius, L. 1999a. 'Differentiation, leaders, and fairness: negotiating climate commitments in the European Community', *International Negotiation*, **4**, 133–66.

Ringius, L. 1999b. *The European Community and climate protection: what's behind the 'empty rhetoric'?*, Oslo: Center for International Climate and Environmental Research.

Risse-Kappen, T. 1995a. 'Bringing transnational relations back in: introduction'. *In:* Risse-Kappen, T. (ed.), *Bringing Transnational Relations Back in: Non-state Actors, Domestic Structures and International Institutions*, Cambridge: Cambridge University Press.

Risse-Kappen, T. 1995b. 'Structures of governance and transnational relations: what have we learned?' *In:* Risse Kappen, T. (ed.), *Bringing Transnational Relations Back in: Non-state Actors, Domestic Structures and International Institutions*, Cambridge: Cambridge University Press.

Rosenau, J. 2000. 'Change, complexity, and governance in a globalizing space'. *In:* Pieere, J. (ed.), *Debating Governance: Authority, Steering and Democracy*, Oxford: Oxford University Press.

Rosenau, J. 2007. 'Governing the ungovernable: the challenge of a global disaggregation of authority', *Regulation and Governance*, **1,** 88–97.

Rowlands, I. 1995. 'The climate change negotiations: Berlin and beyond', *Journal of Environment and Development*, **4**, 145–63.

Royden, A. 2002. 'US Climate change policy under President Clinton: a look back', *Golden Gate University Law Review*, **32**, 415–78.

Ruggie, J., Katzenstein, P., Keohane, R. and Schmitter, P. 2005. 'Transformations in world politics: the intellectual contributions of Ernst B. Haas', *Annual Review of Political Science*, **8**, 271–96.

Schattschneider, E.E. 1960. *The Semi-soveriegn People: A Realist's View of Democracy in America*, New York: Holt, Rinehart and Winston.

Schelling, T. 1960. *The Strategy of Conflict*, Cambridge: Harvard University Press.

Schleich, J., Eichhammer, W., Boede, U., Gagelmann, F., Jochem, E., Schlomann, B. and Ziesing, H.-J. 2001. 'Greenhouse gas reductions in Germany – lucky strike or hard work?', *Climate Policy*, **1**, 363–80.

Schroeder, M. 2008. 'Transnational NGO cooperation for China's climate politics', *Cambridge Review of International Affairs*, **21**, 505–25.

Sebenius, J. 1984. *Negotiating the Law of the Sea*, Cambridge: Harvard University Press.

Sebenius, J. 1994. 'Towards a winning climate coalition'. *In:* Mintzer, I. and Leonard, A. (eds), *Negotiating Climate Change: The Inside Story of the Rio Convention*, Cambridge: Cambridge University Press.

Sell, S. 2003. *Private Power, Public Law: The Globalization of Intellectual Property Rights*, Cambridge: Cambridge University Press.

Sewell, G. 1996. 'Conflicting beliefs: national implementation of the United Nations' framework convention on climate change', *Environmental Impact Assessment Review*, **16**, 137–50.

Sewell, G. 2005. *Actors, Coalitions and The Framework Convention on Climate Change*, Massachusetts: Massachusetts Institute of Technology.

Shabecoff, P. 1988. 'Global warming has begun, expert tells senate', *New York Times*, June 24.

Silverman, D. 2001. *Interpreting Qualitative Data: Methods for Analysing Talk, Text and Interaction*, London: Sage Publications.

Skjaerseth, J.B. and Wettestad, J. 2008. *EU Emissions Trading: Initiation, Decision-making and Implementation*, Aldershot: Ashgate.

Slaughter, A.-M. 2004. *A New World Order*, Princeton: Princeton University Press.

Smith, J. and Wiest, D. 2005. 'The uneven geography of global civil society: national and global influences on transnational association', *Social Forces*, **84**, 621–39.

Spector, B. 2003. 'Deconstructing the negotiations of regime dynamics'. *In:* Spector, B. and Zartman, I.W. (eds), *Getting It Done: Postagreement Negotiation and International Regimes*, Washington DC: United States Institute of Peace Press.

Spector, B. and Zartman, W. 2003. 'Regimes and negotiation: an introduction'. *In:* Spector, B. and Zartman, W. (eds), *Getting It Done: Postagreement Negotiation and International Regimes*, Washington DC: United States Institute of Peace Press.

Stern, N. 2007. *The Economics of Climate Change: The Stern Review*, Cambridge, Cambridge University Press.

Stern, T. 1999. *Press briefing by Todd Stern, US President's Coordinator for Climate Change, 3 June* [Online]. http://www.presidency.ucsb.edu/ws/index.php?pid=47854#. 2009. (accessed 17 September 2013).

Stiglitz, J. 2006. *Making Globalization Work*, Camberwell: Allen Lane.

Stokke, O. 2004. 'Trade measures and climate compliance: institutional interplay between WTO and the Marrakesh Accords', *International Environmental Agreements: Politics, Law, and Economics*, **4**, 339–57.

Susskind, L. 1994. *Environmental Diplomacy: Negotiating More Effective Global Agreements*, New York: Oxford University Press.

Susskind, L. and Crump, L. (eds) 2008. *Multiparty Negotiation*, Los Angeles: Sage Publications.

Tarrow, S. 2005. *The New Transnational Activism*, New York: Cambridge University Press.

Thatcher, M. 1989. 'Speech to United Nations General Assembly (Global Environment)', New York: United Nations.

The Economist. 1992. 'Europe's industries play dirty', *The Economist*, 9 May, p. 7758.

Toman, M. 2004. 'Economic analysis and the formulation of US climate policy'. *In:* Lutter, R. and Shogren, J. (eds) *Painting the White House Green: Rationalizing Environmental Policy Inside the Executive Office of the President*, Washington DC: RFF Press.

True, J. and Mintrom, M. 2001. 'Transnational networks and policy diffusion: the case of gender mainstreaming', *International Studies Quarterly*, **45**, 27–57.

Trumbore, P. 1998. 'Public opinion as a domestic constraint in international negotiations: two-level games in the Anglo-Irish peace process', *International Studies Quarterly*, **42**, 545–65.

UNFCCC 1992. 'United Nations framework convention on climate change', New York: United Nations.

UNFCCC 1995a. 'First conference of the parties: directory of participants'. *FCCC/1995/Inf.5/Rev.2.*

UNFCCC 1995b. 'Matters relating to commitments. Review of the adequacy of Article 4, paragraph 2 (A) and (B). Comments from Parties and other member States'. *FCCC/CP/1995/Misc.1.*

UNFCCC 1995c. 'Report of the conference of the parties on its first session, held at Berlin from 28 March to 7 April'. *FCCC/CP/1995/7/Add.1.*

UNFCCC 1997a. 'Kyoto Protocol to the United Nations framework convention on climate change'. New York: United Nations.

UNFCCC 1997b. 'Third conference of the parties: directory of participants'. *FCCC/1997/Inf.5.*

UNFCCC 2000. 'Methodological issues land-use, land-use change and forestry (submissions from parties) addendum'. *FCCC/SBSTA/2000/MISC.6/Add.1.* Bonn: UN Framework Convention on Climate Change Secretariat 4 September 2000.

United States Senate 1997a. 'Congressional record', *Volume 143*, Number 89.

United States Senate 1997b. 'Senate resolution 98', *105th Congress*, 1st Session.

US Department of State 2000. 'Compliance-related issues: submission of the United States', *Submitted to the UN Framework Convention on Climate Change (UNFCCC) Secretariat, February 4.*

US Government 1993. 'Climate change action plan', Washington DC: White House.

van Asselt, H. 2010. 'Emissions trading: the enthusiastic adoption of an "alien" instrument?'. *In:* Jordan, A., Huitema, D., Asselt, H.V., Rayner, T. and Berkhout, F. (eds) *Climate Change Policy in the European Union: Confronting the Dilemmas of Mitigation and Adaptation?*, Cambridge: Cambridge University Press.

Victor, D. and Salt, J. 1994. 'From Rio to Berlin: managing climate change', *Environment*, **36**, 6–32.

Waltz, K. 1979. *Theory of International Politics*, New York: Random House.

Wapner, P. 1995. 'Politics beyond the state: environmental activism and world civic politics', *World Politics*, **47**, 311–40.

Weale, A., Pridham, G., Cini, M., Konstadakopulos, D., Porter, M. and Flynn, B. 2000. *Environmental Governance in Europe: An Ever Closer Ecological Union?*, New York: Oxford University Press.

White, R. 1998. 'Kyoto and beyond', *Issues in Science and Technology*, **14**, 59–65.

Wohlforth, W. 2008. Realism. *In:* Reus-Smit, C. and Snidal, D. (eds), *The Oxford Handbook of International Relations*, Oxford: Oxford University Press.

WWF 1996. 'Policies and measures to reduce CO_2 emissions by efficiency and renewables. A preliminary survey for the period to 2005', The Netherlands: Utrecht University.

Yamin, F. and Depledge, J. 2004. *The International Climate Change Regime: A Guide to Rules, Institutions and Procedures*, Cambridge: Cambridge University Press.

Yellen, J., Testimony of Dr Janet Yellen, Chair, Council of Economic Advisors. House Commerce Committee on the Economics of the Kyoto Protocol, March 4, 1998, Washington DC.

Yin, R. 2009. *Case Study Research Design and Methods*, London: Sage Publications.

Young, O. 1999. 'Regime effectiveness: taking stock'. *In:* Young, O. (ed.) *The Effectiveness of International Environmental Regimes*. Cambridge: The MIT Press.

Young, O. and Levy, M. 1999. 'The effectiveness of international environmental regimes'. *In:* Young, O. (ed.) *The Effectiveness of International Environmental Regimes*, Cambridge: The MIT Press.

Zartman, I.W. 1989. 'Prenegotiation: phases and functions'. *International Journal*, **44,** 237–53.

Zartman, W. 2003. Negotiating the rapids: the dynamics of regime formation. *In:* Spector, B. and Zartman, W. (eds) *Getting It Done: Postagreement Negotiation and International Regimes*, Washington DC: United States Institute of Peace Press.

Zartman, I.W. and Berman, M. 1982. *The Practical Negotiator*, New Haven; Yale University Press.

Zhang, Z. 2009. 'Multilateral trade measures in a post-2012 climate change regime? What can be taken from the Montreal Protocol and the WTO?', *Energy Policy*, **37,** 5105–12.

Index